W9-AVR-944

382.41372 ROSE 2010
Rose, Sarah
For all the tea in China

Central   01/20/2011

CENTRAL LIBRARY

For All the Tea in China

For All the Tea in China

HOW ENGLAND STOLE
THE WORLD'S FAVORITE DRINK
AND CHANGED HISTORY

Sarah Rose

VIKING

VIKING

Published by the Penguin Group

Penguin Group (USA) Inc., 375 Hudson Street, New York, New York 10014, U.S.A.
Penguin Group (Canada), 90 Eglinton Avenue East, Suite 700, Toronto, Ontario,
Canada M4P 2Y3 (a division of Pearson Penguin Canada Inc.)
Penguin Books Ltd, 80 Strand, London WC2R 0RL, England
Penguin Ireland, 25 St. Stephen's Green, Dublin 2, Ireland
(a division of Penguin Books Ltd)
Penguin Books Australia Ltd, 250 Camberwell Road, Camberwell, Victoria 3124, Australia
(a division of Pearson Australia Group Pty Ltd)
Penguin Books India Pvt Ltd, 11 Community Centre, Panchsheel Park,
New Delhi – 110 017, India
Penguin Group (NZ), 67 Apollo Drive, Rosedale, North Shore 0632, New Zealand
(a division of Pearson New Zealand Ltd)
Penguin Books (South Africa) (Pty) Ltd, 24 Sturdee Avenue, Rosebank,
Johannesburg 2196, South Africa

Penguin Books Ltd, Registered Offices: 80 Strand, London WC2R 0RL, England

First published in 2010 by Viking Penguin, a member of Penguin Group (USA) Inc.

1 3 5 7 9 10 8 6 4 2

Copyright © Sarah Rose, 2010
All rights reserved

LIBRARY OF CONGRESS CATALOGING IN PUBLICATION DATA
Rose, Sarah.
For all the tea in China : how England stole the world's favorite drink and changed
history / Sarah Rose.
p. cm.
First published: London : Hutchinson, 2009, with title For all the tea in China : espionage,
empire, and the secret formula for the world's favourite drink.
Includes bibliographical references and index.
ISBN 978-0-670-02152-9
1. Tea trade—Great Britain—History—19th century. 2. Tea trade—China—History—19th
century. 3. Tea—Great Britain—History—19th century. 4. Tea—China—History—19th
century. 5. Fortune, Robert, 1813–1880—Travel—China. 6. Spies—Great Britain—Biography.
7. Business intelligence—Great Britain—History—19th century. 8. East India Company—
History—19th century. 9. China—Description and travel. 10. Himalaya Mountains—
Description and travel. I. Title.
HD9198.G72R67 2010
382'.413720941—dc22   2009041482

Printed in the United States of America
Designed by Nancy Resnick
Title page illustration © Dorling Kindersley

Without limiting the rights under copyright reserved above, no part of this publication may be
reproduced, stored in or introduced into a retrieval system, or transmitted, in any form or by any
means (electronic, mechanical, photocopying, recording or otherwise), without the prior written
permission of both the copyright owner and the above publisher of this book.

The scanning, uploading, and distribution of this book via the Internet or via any other means
without the permission of the publisher is illegal and punishable by law. Please purchase only
authorized electronic editions and do not participate in or encourage electronic piracy of copy-
rightable materials. Your support of the author's rights is appreciated.

For Scott

The greatest service which can be rendered to any country is to add a useful plant to its culture.

—Thomas Jefferson

[Tea] is an exceedingly useful plant; cultivate it, and the benefit will be widely spread; drink it, and the animal spirits will be lively and clear.

—Robert Fortune, quoting a Chinese proverb

Contents

Prologue

There was a time when maps of the world were redrawn in the name of plants, when two empires, Britain and China, went to war over two flowers: the poppy and the camellia.

The poppy, *Papaver somniferum*, was processed into opium, a narcotic used widely throughout the Orient in the eighteenth and nineteenth centuries. The drug was grown and manufactured in India, a subcontinent of princely states united under the banner of Great Britain in 1757. Opium was marketed, solely and exclusively, under the aegis of England's empire in India by the Honourable East India Company.

The camellia, *Camellia sinensis*, is also known as tea. The empire of China had a near complete monopoly on tea, as it was the only country to grow, pick, process, cook, and in all other ways manufacture, wholesale, and export "the liquid jade."

For nearly two hundred years the East India Company sold opium to China and bought tea with the proceeds. China, in turn, bought opium from British traders out of India and paid for the drug with the silver profits from tea.

The opium-for-tea exchange was not merely profitable to England but had become an indispensable element of the economy.

Nearly £1 in every £10 sterling collected by the government came from taxes on the import and sale of tea—about a pound per person per year. Tea taxes funded railways, roads, and civil service salaries, among the many other necessities of an emergent industrial nation. Opium was equally significant to the British economy, for it financed the management of India—the shining jewel in Queen Victoria's imperial crown. While it had always been hoped that India would become economically self-sustaining, by the mid-nineteenth century England was waging a series of expansionist wars on India's North-West Frontier that were swiftly draining whatever profits could be derived from the rich and vast subcontinent.

The triangular trade in botanical products was the engine that powered a world economy, and the wheels of empire turned on the growth, processing, and sale of plant life: poppies from India and camellias from China, with a cut from each for Great Britain.

By the middle of the nineteenth century the British-Chinese relationship was a tragically unhappy. The Exalted and Celestial Emperor in Peking had "officially" banned the sale of opium in China in 1729, but it continued to be smuggled in for generations afterward. (Notably, the sale of opium was also forbidden by Queen Victoria within the British Isles. She, however, was largely obeyed.) Opium sales increased quickly and steadily; there was a fivefold growth in volume in the years 1822–37 alone. Finally, in 1839, the leading Chinese court official in the trading port of Canton, rankled by the profligacy of the foreigners and the pestilence of opium addiction among his own people, held the entire foreign encampment hostage, ransoming the three hundred Britons for their opium, then worth $6 million (about $145 million in today's dollars). When the opium was surrendered and the hostages released, the mandarin ordered five hundred Chinese coolies to foul nearly three million pounds of the drug with salt and lime and then wash the mixture out into the Pearl River. In response, young Victoria sent Britain's navy to war to keep the lucrative opium-for-tea arrangement alive.

In battle, Britain trounced China, whose rough wooden sailing junks were no match for Her Majesty's steam-powered modern navy. As part of the peace treaty, England won concessions from the Chinese that after a century of diplomatic entreaty no one had thought possible: the island of Hong Kong plus the cession of five new treaty or trading ports on the mainland.

Few Westerners had penetrated the Chinese interior since the days of Marco Polo. For two hundred years prior to the First Opium War, British ships had been restricted to docking at the entrepôt of Canton, a southern trading city at the mouth of the Pearl River. Britons could not officially step foot outside their warehouses, and many had never even seen the city walls, 20 feet thick and 25 feet high and only 200 yards away from the foreigners' district. Now, with their triumph in the war, the interior of China was opened to the British—just a crack—for business.

With five new cities in which to trade, British merchants began dreaming of the lush silks, delicate porcelains, and perfumed teas stockpiled in the Chinese interior, just waiting to be sold to the wider world. Merchants conceived of the possibility of dealing directly with Chinese manufacturers, rather than the cantankerous middlemen or Hongs who commanded the Canton warehouses. Bankers had visions of untold riches, of mineral wealth, and of crops, plants, and flowers—a giant country filled with unmonetized commodities.

The new order established by the First Opium War was an unstable one, however. Forced by British gunboats to sign the intolerable treaties, China, the once proud and self-contained nation, had been thoroughly shamed. British politicians and traders worried that the humiliated Chinese emperor might upset the delicate balance established by the peace accords by legalizing opium production in China itself, thus breaking India's (and, in turn, Britain's) monopoly on the poppy.

An idea now took hold in the City of London: Tea could and must be secured for England. The Napoleonic Wars had long since ended by the time of the Opium Wars, but the brave men who had fought at Trafalgar and Waterloo still held enormous sway over foreign policy and opinion. Henry Hardinge, a great general who had helped defeat Napoleon beside Lord Nelson and the Duke of Wellington, warned of the risk posed by a defiant China when he was governor-general of India:

> It is in my opinion by no means improbable that in a few years the Government of Pekin, by legalising the cultivation of Opium in China, where the soil has been already proved equally well adapted with India to the growth of the plant, may deprive this Government of one of its present chief sources of revenue. Under this view I deem it most desirable to afford every encouragement to the cultivation of Tea in India; in my opinion the latter is likely in course of time to prove an equally prolific and more safe source of revenue to the state than that now derived from the monopoly on Opium.

If China legalized the opium poppy, it would leave a crucial gap in the economic triangle: England would no longer have the money to pay for her tea, her wars in India, or her public works projects at home. Chinese-grown opium would put an end to the shameful economic codependence between the two empires, the unhappy marriage sealed by the exchange of two flowers. It was a divorce that Britain could ill afford.

The Indian Himalaya mountain range resembled China's best tea-growing regions. The Himalayas were high in altitude, richly soiled, and clouded in mists that would both water tea plants and shade them from the scorching sun. Frequent frosts would help sweeten and flavor their liquor, making it more complicated, intense, delicious.

As botanical products swelled the balance sheets of the Oriental trade, they became so important to the world order that the men who studied them—men who were once popularly regarded as mere gardeners—began to be appreciated as the botanists they were. By the mid-nineteenth century, botanists were no longer viewed as humble men in hats and hobnailed boots, tending to their bulbs, flowers, and shrubs, but as swashbucklers and world-changers, whose collections of foreign plants had potential scientific, economic, and agricultural value in England and throughout the empire. New technology for transplanting live flora had also grown more sophisticated, allowing professional plant hunters to collect and transport ever more exotic specimens.

No longer confined to China's southernmost coast, Britain now had greater access to the areas where tea was cultivated and processed. If the manufacture of tea in India was to be successful, Britain would need healthy specimens of the finest tea plants, seeds by the thousand, and the centuries-old knowledge of accomplished Chinese tea manufacturers. The task required a plant hunter, a gardener, a thief, a spy.

The man Britain needed was named Robert Fortune.

I

Min River, China, 1845

On an autumn afternoon in 1845, long before Robert Fortune made his name as one of the world's great plant hunters, it seemed very likely that he would die in China. For two weeks he had been confined to a listless junk near the city of Fuzhou, at the mouth of the Min River. His ordinarily robust constitution was near collapse. With a raging fever, he took to his bunk in the cabin of a seagoing cargo boat, dizzy from the smell of bilgewater and rotting fish. The junk's deck, laden with timber from the countryside, also held Fortune's cargo, which included trunk-size glass boxes filled with flowers, shrubs, grasses, vegetables, fruits, and all manner of exotic plant life. These glazed cases, known as Wardian or Ward's cases after their inventor, were on their way with Fortune to London—if he ever made it that far. With his long legs dangling off a bunk made for the shorter Chinese, Fortune, only thirty-three years old at the time, imagined himself dying in the boat's hold, being swaddled in his grimy bedclothes, and then unceremoniously hurled overboard into a watery grave.

He was in the last days of a three-year expedition to China, conducted at the behest of the Royal Horticultural Society of

London, to find and collect samples of the Orient's botanical treasures. Fortune's assignment included procuring "the peaches of Pekin, cultivated in the Emperor's Garden and weighing 2 lbs.," among other imagined delicacies. In addition to living flora, he would take home a pressed herbarium and intricate botanical drawings penned by the finest draftsmen in China. With every seed, plant, graft, and clone collected, Fortune was advancing Western knowledge of the East and of botany.

Each new plant he cataloged was significant not only for its novelty value but for its possible utility to the British Empire. The nineteenth-century world had been revolutionized by the mechanized manufacture of natural products into refined goods: Cotton became cloth on automated looms, iron ore was transformed into train tracks and steamships' hulls, and clay became stoneware and porcelain. China was a frontier of agricultural riches and industrial possibilities.

But, lying feverish in his bunk, Fortune could not believe he or his plants would ever find their way safely back to the gardens of Britain. He was in the greatest danger he had ever known, notwithstanding the three years he had already spent living in China as a foreigner.

"It seemed hard for me to die . . . without a friend or countryman to close my eyes, or follow me to my last resting place; home, friends, and country, how doubly dear they did seem to me then!" he later wrote.

Fortune's life was emblematic of that of many entrepreneurial Britons who were seizing the opportunities offered by the expansion of the empire. His roots were modest. His early education in rural horticulture took place at the elbow of his "hedger" father, a hired farmworker. He had no formal higher education beyond his parish schooling in the tiny town of Edrom, in the Scottish

Borders; his knowledge of natural history was not obtained at the universities of Oxford or Edinburgh, but from folk practice and professional apprenticeship. He gained a first-class certificate in horticulture, a trade qualification, but lacked the degree in medicine that was a common accompaniment to an interest in botany among those he aspired to join as a peer. For all that, Fortune was ambitious, and for many nineteenth-century Scots as well as English second sons of some talent and no sinecure, seeking one's chances abroad was the only way to advance in the rigid Victorian social hierarchy. There were untold possibilities to make a decent living by exploiting the untapped resources of the empire.

With his lively mind, Fortune rose quickly through the ranks of horticulture, first in the Botanic Garden at Edinburgh and later at the Royal Horticultural Society's gardens in Chiswick. Based on his skills at cultivating orchids and hothouse ornamentals—the rare, showy plants from the Orient—Fortune was the Society's first choice to be dispatched to explore China at the close of the First Opium War. Founded in 1804 by John Wedgwood, Charles Darwin's maternal uncle, the Society was the arbiter of all things green and growing in England. At its lively meetings botanists and zoologists presented papers and discussed the latest developments in their fields, which were multiplying rapidly with the increase in British global dominion. Its journals detailed the classification of the newest plants from the farthest reaches of Her Majesty's empire. The Society's botanists were busily engaged in the great project of naming and describing every species according to the methods by which they reproduced, a system recently introduced in Europe by the great Carl von Linné, known as Linnaeus.

Victorian England had a passion for natural exotica, for the insects, fossils, and plants that had been collected over the decades by its missionaries, officers, and merchants on the high seas. With the rural peasantry moving off the land and into the city as industrialization took hold and farmland was enclosed by the gentry, Britons began to yearn for nature in all its forms, and a new market evolved to supply British households with plants. Potted ferns of all varieties became a national obsession and soon were seen everywhere: on decorative porcelain, wallpaper, and textiles; in the conservatories of the rich; and on the windowsills of the poor. Easy to grow and propagate, and hardy enough to transplant, ferns had a wild, uncultivated look that reflected the national pastoralism.

In pursuit of a more exotic trophy, in 1856 the Sixth Duke of Devonshire paid a hundred guineas (about $12,000 today) for the first imported example of the Philippine orchid *Phalaenopsis amabilis*. The duke nearly squandered his fortune on his passion for flowers. Striking and delicate with its snow-white oblong petals and yellow lips, the amabilis orchid was beloved by Society members and hugely profitable to its discoverer.

Since China had been closed to Westerners for centuries, it remained largely a blank on plant-hunting maps, a place once marked "Here be dragons." China considered itself the center of the world, but the Middle Kingdom was in fact almost entirely removed from the global stage, albeit Chinese civilization was more than five thousand years old. With little real knowledge of China, the West projected onto it a million fantasies of paradise, danger, and exoticism. All the yearning reflected in England's desire for gardens was fulfilled and even magnified by its perception of China as the untouched Shangri-la of horticulture.

Had Europeans been permitted a closer look at China, they would have found a country riven by internal unrest and governed by hated foreigners. The Manchus had crossed the Great Wall from the north and for two hundred years had ruled the ethnic Han Chinese from the capital of Peking (now Beijing), demanding fealty and taxes. Secret societies abounded in the south, sworn to combat the alien emperors of the Manchurian Qing Dynasty. The countryside was full of thieves and highwaymen, and the seaways were plagued by pirates. Famines blighted the lives of the rural peasantry, as did corrupt officials, members of the Confucian-educated scholar class known as mandarins, while squalor overwhelmed the cities.

The British had some knowledge of China's affairs through trading contact—the East India Company had been doing business in Canton for almost two centuries—but the interior of China was for the most part terra incognita. Two things England did know, however: There were bound to be marvelous plants in China, and the economic future of Great Britain might benefit greatly from them.

The emperor of China took pains to prevent interlopers from exploring his territory and capitalizing on its resources. In the wake of the First Opium War, the Treaty of Nanking had granted Britain rights to trade in Fuzhou and four other treaty ports, walled coastal cities previously forbidden to outsiders. But suspicion of the British and their intentions remained widespread, with white men officially prohibited from traveling beyond the newly constructed foreign concessions in the port cities. And if Chinese laws could not keep foreigners inside the city walls, the realities of life in China might: It presented a hostile environment to Britons unaccustomed to the humidity, insects, vermin, disease, and terrible sanitation of even the most civilized outposts. No sane man would wish to live or die in China.

In the autumn of 1842, news of the peace between China and England reached the halls of the Royal Horticultural Society, which provided it an unprecedented opportunity to send an expedition into deepest China. The discovery and exploitation of botanical materials were now widely recognized as a British priority, and Robert Fortune was the first person given leave by the Foreign Office to travel to China at the end of hostilities.

Fortune was chosen for the China expedition despite lacking the usual gentleman's background that would fit him for such a prestigious Society assignment. He was paid wages of only £100 a year (about £5,000 in today's money, or $10,000), a paltry sum on which to raise a family and one that would not be increased during his entire three-year tenure. When he dared to try to negotiate for a better stipend, the Society sharply rebuked him, with a reminder that "the mere pecuniary returns of your mission ought to be but a secondary consideration" next to "the distinction and status which you could not have attained any other way."

Given his social standing and lack of property, Fortune was not judged by the Society as being entitled to any of their perquisites, including such niceties as a rifle, pistols, bullets, and gunpowder. His mission was to study and expropriate the rare plants of the Orient—a task that did not, they maintained, require weaponry. It was not for the plants, Fortune argued, but for his own safety that he needed this protection. While his fellow professional botanists were sympathetic, the fact remained that when gentlemen plant hunters needed guns, they had the independent means with which to purchase them.

The Society members eventually agreed that their investment

in the China expedition would be forfeited or at least greatly reduced if Fortune were killed before its completion. While they
again refused to raise his salary, they did reluctantly provide him
with some weapons.

As it turned out, their choice of Fortune to lead the expedition
was a triumph. He faithfully reported back to them all the details
of his botanical finds, shipping to England as many living examples of the new and wonderful plants of China as he could. He
took cuttings, made grafts, kept careful notes, and wrote detailed
letters for the botanists of the world, many of whom received the
fruits of his discoveries as part of a program of global imperial
plant exchange. By the end of his first trip to China, Fortune was
considered a success within the circles of scientific exploration,
the first of his shipments having been received, propagated, and
duly celebrated.

It helped that he had a collector's eye for the rare and the beautiful, as well as for potential market value. He plumbed the reaches
of China's natural wonders, keeping his eye trained on the flowers
that, while perhaps not of immense significance to science, might
nonetheless fetch a high price under the gavel. Over the course of
three years Fortune discovered the winter-blooming jasmine; the
bleeding heart, a floral image of brokenheartedness to play to Victorian romanticism; the Chinese fan palm, a gift of colonial exotica to Queen Victoria on her thirty-second birthday; the white
wisteria; the corsage gardenia (*Gardenia fortunei*); and the lilac
daphne (*Daphne fortunei*). Fortune found the fabled double yellow
tea rose (commonly known as Fortune's Double Yellow or The
Gold of Ophir) in a mandarin's garden, climbing walls to a height
of 15 feet. One discovery in particular—the kumquat, *Citrus fortunei*, or commonly *Fortunella*, the miniature citrus fruit with edible
skin—would make him immortal. Although Fortune did not own

the property rights to any of his botanical finds, he would travel home with many other salable curiosities and trinkets, rare gems, pottery, and pieces of jade.

Beyond his scrupulously observed field notes, Fortune kept a diary of his exploits and of his encounters with the exotic people and customs of China. He wrote about his servants and interpreters, officials, merchants, herbalists, artists, fishermen, gardeners, monks, prostitutes, street peddlers, women, and children. Like other travelers of the Victorian era, upon his return in 1847 he published this document in the form of a travelogue. *Three Years' Wanderings in the Northern Provinces of China* is liberally sprinkled with the geographical and botanical descriptions one might find in any horticultural study, but it also contains its author's jaunty and unstinting reminiscences of meeting fellow British expatriates in the treaty ports, of the temples and priests, and of the dangers of bandits.

Fortune's trip began in Hong Kong, Britain's newest colonial possession, during the typhoon season of 1843. He declared the island to be "in lamentable condition," suffering from the bad air, or *mal aria*, which laid waste its European inhabitants. "Viewed as a place of trade, I fear Hong Kong will be a failure," he wrote none too prophetically. Sailing up the coast toward the northernmost trading port at Shanghai, Fortune was nearly shipwrecked in a typhoon. "Some idea may be formed of the storm when I mention that a large fish weighing at least 30 pounds was thrown out of the sea onto the skylight upon the poop, the frame of which was dashed to pieces and the fish fell through and landed upon the cabin table." While plant hunting on mainland hillsides, Fortune was pickpocketed, chased, and beaten by thieves, who threw a brick at his head. "I was stunned for a few seconds and leaned against a wall to breathe and recover myself. . . . The rascals again surrounded me and relieved me of several articles," he wrote.

Fortune also visited opium-trading dens and held forth on the perils of addiction. "I have often seen the drug used and I can assert in the great majority of cases it is not immoderately indulged in. At the same time, I am well aware that, like the use of ardent spirits in our own country, it is frequently carried to a most lamentable excess."

He was particularly eager to obtain plant material from the gardens of mandarins, which often held some of the best specimens available. In order to access the gardens of Suzhou, a forbidden city, he donned a disguise. "I was, of course, travelling in the Chinese costume; my head was shaved, I had a splendid wig and tail, of which some Chinaman in former days must have been extremely vain; and upon the whole I believe I made a very fair Chinaman." The masquerade fooled the gatekeepers, and Fortune observed, "How surprised they would have been had it been whispered that an Englishman was standing amongst them."

Three Years also charted Fortune's evolution as a man gradually coming to terms with what he viewed as an enigmatic society. At first he approached China with all the arrogance of a colonialist, dismissing it as a country full of "wretched Chinese hovels, cotton fields, and tombs." Like many foreigners, he saw himself as a missionary for the Western way of life, mocking any Chinese notions of superiority. European expatriates should, he believed, serve as examples to the Chinese, since any "peeps at our comforts and refinements may have a tendency to raise the 'barbarian race' a step or two higher in the eyes of the 'enlightened' Chinese." Yet after three years his opinion had been tempered, for he could not have successfully completed his mission without coming into close contact with everyday Chinese, and in so doing the country inevitably began to assume a human face for him.

The barriers to penetrating China were enormous and ranged

from the linguistic and the social to the strictly official. In light of his outsider status Fortune was almost entirely dependent on Chinese peasants, boatmen, coolies, guides, and porters. He found many who were willing to help him sidestep national and cultural boundaries—for a price. The contact he had with ordinary Chinese people, which few Westerners had previously experienced, led him to hope for unity and reconciliation between the two nations. "Nothing," he wrote, "can give the Chinese a higher idea of our civilisation and attainments than our love for flowers, or tend more to create a feeling between us and them."

Three Years was both a critical and a popular triumph. A reviewer for *The Times* of London, the paper of record, wrote as follows:

> When readers have recovered from the intoxication produced by the exciting drink of Uncle Tom's Cabin, we seriously recommend them . . . to "try Fortune's mild Bohea." There is no adulteration in the article. It is pure—almost to a fault, and has to be taken, as the Chinese themselves drink tea, without the admixture of milk and sugar, for luscious ornament and superfluous additions our singleminded author has none. Concerning the flavour there can be no mistake. One trial will prove the excellence of the commodity, and he that sips once will be soothed and sip again.

The book was avidly consumed by an audience of armchair botanists and starry-eyed colonialists, but also by those who simply enjoyed a gripping tale. Fortune rendered his experiences in the traditional form of the Victorian bildungsroman: Living by his wits and full of improvisation, the young man from the provinces made his name in the unlikeliest of ways and was celebrated upon his return to London.

But if one vignette in particular stood out and established Fortune in the popular imagination as a hero, it was the tale that began with him in his fevered state, lying in a cabin belowdecks on the treacherous Min River.

Fortune's flat-bottomed ship made a turn north from the rocky mouth of the river out into the South China Sea. The small wooden craft sailed on an early morning wind, her rattan sail full, patched together like a quilt and pouching out between bamboo stays.

The cabin door burst open, and the breathless captain and pilot began to shout in Fukienese, the dialect of the coast.

"Pirates!" they warned. *Haidao!*

On deck it was a scene of chaos, as the captain had begun ripping up floorboards in order to stash his valuables while crew members likewise sought places to hide the few pieces of copper money they had managed to set aside in a life of hard sea labor. Fortune took out his telescope and could see five ships on the horizon, unmarked, flying no imperial flags. They could only be pirate ships.

As the first enemy craft bore down on them, its crew, some fifty or so strong, gathered along the gangway and began "hooting and yelling like demons."

"Their fearful yells seem to be ringing in my ears even now, after this lapse of time, and when I am on the other side of the globe," Fortune later wrote.

Turning back to the crew of his own ship, he noticed they had been subtly transformed. The men had made themselves look like beggars, as if they had been at sea for forty years or more; they were now wearing only the suggestion of clothes—torn rags and shreds of rice sacks. With enemies about to board, there was nothing to be gained by appearing prosperous.

Piracy was the scourge of China. Trade with the West brought in untold foreign wealth, leaving the coasts a battleground between the mandarins, who sought to impose official control over shipping, and a mutinous water world of thieves. If life in a pirate gang was a frenzy of sodomy, gang rape, torture, and cannibalism, it was also a way to make a living outside the rigid structure of Qing society. There was little profit in taking a boat as small as Fortune's, as it would yield only a few captives to sell as slaves and a negligible vessel to commandeer or sink—but there was also very little risk. The pirate ship could easily outgun, outman, outmaneuver, and generally outclass Fortune's cumbersome cargo junk.

His crew now began hauling up baskets of stones from the hold, emptying the rocks across the span of the deck. In peacetime these stones functioned as ballast; in war they were the most rudimentary of weapons. However, as Fortune noted, "All the pirate junks carried guns, and consequently a whole deck load of stones could be of very little use."

"Bring the junk around," one of the crew demanded.

"Run us back to the cliffs and hide among them," another argued.

"Fight!" cried one. *Da!*

"Flee!" yelled another. *Zou!*

The fate of a Westerner taken by pirates was often a bloody one: The brigands would possibly hold Fortune for ransom and torture him. Under duress he would be forced to write letters to the British missions, begging for impossible sums to be paid to secure his release. One English captive held in such circumstances wrote, "I saw one man nailed through his feet with large

nails, then beaten with four rattans twisted together, till he vom-
ited blood; and after remaining some time in this state, he was
taken ashore and cut into pieces." Another "was fixed upright, his
bowels cut open and his heart taken out, which they afterwards
soaked in spirits and ate."

Pirates did not shrink from such cruelty, given that their own
death sentences were virtually assured. The state's punishment
for piracy was nailing the perpetrators to a cross, slicing them
with a sharp knife, and cutting them into 120, 72, 36, or 24 pieces.
Of the lightest sentence, 24 cuts, one observer wrote:

> The first and second cuts remove the eyebrows; the
> third and fourth, the shoulders; the fifth and sixth,
> the breasts, the seventh and eighth, the parts between
> each hand and elbow; the ninth and tenth, the parts
> between each elbow and shoulder; the eleventh and
> twelfth, the flesh of each thigh; the thirteenth and
> fourteenth, the calf of each leg; the fifteenth pierces
> the heart; the sixteenth severs the head from the body;
> the seventeenth and eighteenth cut off the hands; the
> nineteenth and twentieth, the arms; the twenty-first
> and twenty-second, the feet; the twenty-third and
> twenty-fourth, the legs.

Fortune knew that he would be the choicest of the pirates'
prizes, so he mustered his depleted strength, rallied from his
fever, and imposed what order he could on the pandemonium on
deck. Raising his pistol, he took aim at the head of his own
helmsman.

"My gun is nearer to you than those of the pirates," he threatened, "and if you move from the helm, depend upon it, I will shoot you."

As he delivered his warning, the enemy ship fired a broadside cannon. The crew—every man but Fortune's terrified pilot—fled belowdecks. The cannonball whizzed over Fortune's head, its flight taking it directly between the sails.

The pirate ship was at least half again the size of Fortune's, but as her guns were fixed along her gangways (the passageways along either side of the ship), the craft was forced to adjust her head-on course and turn at a sharp angle in order to fire. While the cannon volley continued, Fortune settled on his plan: He would not retaliate from a distance but would allow the pirates to believe his ship could be boarded easily. Bringing them into close range would give the advantage to Fortune's precision weaponry over the pirates' heavy and clumsy cannons and locally made muskets or matchlocks (the Chinese did not have rifles or pistols), which were as likely to blow up in the hands of a marksman as fire a shot.

The pirates bore down on them, guns ablaze. When they were 20 yards off, Fortune took his chance. Crawling along the deck toward the high quarterdeck at the stern, he rose abruptly and let loose with both barrels of his rifle.

In an instant the shocked crew of the attacking ship disappeared behind its bulwarks. Fortune's shot was true, leaving one of the pirates injured and probably dead. The pirate craft was suddenly a ghost vessel: No one was steering her; her sails luffed helplessly.

Fortune's craft, on the other hand, had its pilot and a full sail.

Pirates traveled in packs, however, and shortly afterward a

second ship began to gain on them, and three more were sighted in the distance.

Fortune then hit upon an idea prompted by his rag-bedecked crewmen and his own experiences visiting the mandarin gardens of Suzhou in disguise: cultural cross-dressing in borrowed clothes. He still had several changes of Western clothing stored in his cabin below. What if he dressed up the crew as Occidentals? "It now struck me that perhaps I might be able to deceive the pirates with regard to our strength," he wrote. The Chinese were essentially blind at sea: While all British ships carried telescopes and some even binoculars, few Chinese vessels had them. If the pirates believed there was a full European contingent and arsenal aboard, they might be less enthusiastic about running down the junk.

Fortune dressed the "least Chinese-looking Chinamen" from the crew in his remaining Western finery. In frock coats, trousers, and heeled shoes, the men began to look like Victorian travelers. He instructed them to take up sticks that might look like rifles at long range, and several brandished the short levers used for hoisting the sails.

However convincingly the crew may have appeared British, with pirates at close range they remained petrified. When a cannon volley began, every man fled under the decks, leaving Fortune alone.

The second pirate ship began to fire: more shrapnel, more terror, more cries. But before the brigands had time to reload, Fortune was on his feet.

He let loose the contents of his rifle—two shots, fore and aft—and then fired his revolvers, killing the helmsman and again leaving the vessel to heel under the wind.

On the horizon the remainder of the pirate fleet began to turn about.

Fortune's crew emerged from below, shouting in victory and screaming taunts at the retreating pirates.

"Come back and fight like men!" they called.

The crew picked up the stones littering the decks and pitched them into the water after their retreating enemy.

"A stranger who had not seen these gentry before would have supposed them the bravest men in existence," Fortune later commented. "Fortunately the pirates did not think it proper to accept the challenge."

He wrote, "With the captain, pilot, crew and passengers, I was now one of the greatest and best men in existence. They actually came and knelt before me, as to some superior being."

His seagoing escapade, and the attention it attracted upon publication of his book, helped cement his reputation as a seasoned China hand, revered among the locals in a way that no tradesman, military man, or missionary could ever be. When Fortune returned to China on his next expedition, much more than the fate of a few orchids would be at stake: He would change the fate of nations.

2

East India House, City of London, January 12, 1848

East India House occupied a prestigious site on Leadenhall Street in the center of the City of London. The building's grand façade had Ionic columns supporting a triangular tympanum decorated with emblems of global commerce: At the corners, a figure representing Europe was seated on a horse, and the figure of Asia sat astride a camel; between Europe and Asia rode King George in flowing Roman dress. The mad king, who famously lost the reins of thirteen profitable colonies in the Americas, brandished a sword in defense of international trade. Although the architectural pediment faced north, every man who walked beneath it into the bustling East India Company offices below faced due east, toward the Orient, the center of profit for the venerable company.

Amid the hubbub of the trading house, whose day-to-day activities included the recopying of letters, the distribution of favors, perquisites, and privileges in assemblies that lasted from dawn to dusk, and the serving of breakfast, a wooden chest arrived from India. It was carried through paneled hallways sumptuously adorned with portraits, statues, and memorabilia and beyond a vast library and a museum filled with models, coins, medals, fossils, stuffed birds, sculptures, and reliefs. Shouldered by porters from the dockyards, the chest passed before a clock-work tiger, which, when wound up, would "eat" a wooden British soldier—once the property of an Indian sultan before his defeat at the hands of the company.

Large but light, the chest was delivered to a young clerk, who pried off its tin-lined lid, releasing a fragrant herbal aroma. From the contents of the chest he began to prepare several small pack-ages. With a scale before him and brass weights lined up by size, he measured out uniform quantities from the box, carefully de-positing each into a wax-dipped cloth bag. He was readying par-cels of loose tea to be delivered to the best tea distributors in London.

The task of tea allocation was not among the clerk's typical duties, which were those of any secretary: writing out triplicates of every document, letter, and bill of lading that reached the of-fices of the company from the Orient. To a man making a de-cent wage of £300 a year for work that was neither taxing nor glamorous, this task of doling out some packets of tea would nevertheless be among the most significant actions he would perform in his lifelong career. It was no exaggeration to say that his employer's survival hinged on whether the tea he was dis-patching made a favorable impression on its distinguished recipients.

Officially titled the United Company of Merchants of En-gland Trading to the East Indies, although sometimes referred to as John Company or the Honourable Company, the clerk's em-ployer was a global corporation that had weathered nearly three hundred glorious and mostly profitable years in trade with the East. In that time it colonized much of the world and in the pro-cess became its first and largest multinational company. For very good reason it was called the "Grandest Society of Merchants in the Universe."

When Queen Elizabeth granted her royal charter to the East India Company in 1600, she gave it all trading rights in the East Indies, a mandate as broad and valuable as any public concern has held. For the first hundred years of its existence it largely bought spices and fabrics in the Orient and sold them in London. To fund the expeditions eastward, the company sold shares, and stockholders received a dividend on profits. The operation was enormously successful for England, and the company prospered.

As profits and opportunities grew, however, trade with the East became more and more complicated. John Company became the de facto government of many of the lands in which it did business: It could acquire territory, mint money, command armies, sign treaties, make war and peace, and develop its own judicial and taxation systems. It became a peer to empires and states and, as such, an entirely new entity in the global economy.

The East India Company gave birth to the fortunes of the Pitt family, the military reputation of the Duke of Wellington, and the empire of Hastings. One company governor, Elihu Yale, funded a college of some apparent renown. The clublike offices on Leadenhall Street employed such fine minds as John Stuart Mill and Charles Lamb. With a staff of nearly 350 in the London office, it was the single largest private employer in Britain. The company hired as many soldiers as did the Crown and thus doubled the number of available jobs in the military, while increasing civil service positions by 50 percent. The company extended a gentlemanly capitalism to England's otherwise propertyless leisure class, largely from southern England and public school educated. "Some of the best working blood of England is in India," commented one company man.

Managed in London by its Court of Directors, the East India Company was organized very much along the same lines as modern corporations. Businesses previously had been owned by the same people who managed them. But the shareholders of East India Company had no say in its day-to-day operations. A professional managerial class arose in England, and becoming a member of it became synonymous with middle-class success. The company's international ambitions were so extensive and its structure so complex that it developed elaborate international banking and inventory systems to track goods, services, and debts across oceans. Managers were empowered to make high-volume transactions in several markets simultaneously, using whatever technology and information was available—from the letter to the telegraph to the hunch.

Like today's international businesses, the company would do anything to get ahead of the game, and it was generally believed that tea was the commodity that would keep the company preeminent. Tea was first introduced into England in the 1660s as part of the dowry of Portuguese princess Catherine of Braganza when she married Charles II. Tea proved to be an ideal cargo for the East Indiaman merchant boats in that it was lightweight, packed easily, and could withstand the vagaries of many months spent in an ocean crossing. An exotic luxury, tea rapidly became a favorite way among the upper classes to signify civility and taste in the chilly, wet climate of Britain. From there it rapidly percolated downward through society so that by the mid-eighteenth century tea had become the most popular drink throughout Britain, outselling even beer.

Previously just another trading commodity for the Honourable Company, tea had now become a staple of British life. To be English was to drink tea: Wives put tea on the breakfast table, and the bankers of the empire understood that it was tea that made the Far East trade go round. It was a significant profit center for the government, a multibillion-pound industry, accounting for as much as 10 percent of the total British economy as measured by tax revenues to the exchequer. And the East India Company had its stamp on every single case shipped into England.

The dominance of the company was threatened, however, when in the early nineteenth century Parliament made a series of moves to withdraw its license to trade with Asia. Initially the company's royal charter had given it a total monopoly on trade with the Orient at a time when no one really understood the implications of such a grant. Monopolies, by definition, squelch competition and innovation. Not only did rival firms object to the high barriers to trading in foreign ports, but a growing populism in England extended to the business practices of the empire. As the men of Britain became increasingly enfranchised politically, why couldn't every British trading firm have the same rights to trade in the Far East? In 1813, Westminster withdrew the company's monopoly on trade in India; it was left to rule the subcontinent as the de facto government but was ordered to allow other corporations to do business in Indian ports with Indian-made goods. Despite this check on company power, it collected tax revenues in India equal to half of the total tax revenues of Britain and remained financially healthy. It also retained its most valuable trade monopoly in China.

China had once been a source of pure profit to the company. The company had the run of Cathay, and every chest of tea, silk, or porcelain out of Canton traveled on East Indiaman ships so that money practically minted itself. The margins on tea were particularly high, and its value by the turn of the nineteenth century equaled that of all other Chinese goods combined. But free trade advocates, including Adam Smith, continued to rail against the company's dominance in the China trade. An 1834 act of Parliament finally stripped John Company of its long-sanctioned monopoly over China. Free trade sentiment and mercantilist tendencies stirring in England ensured that there was fierce jostling for a share in the lucrative tea business. New, smaller trading firms were soon docking in Canton, offloading opium, onloading tea for England, and, with their faster ships, sailing between continents in record time. The triple-masted East Indiaman looked old, bloated, and slow in comparison, much like the East India Company itself. It seemed as if the company's days of unrivaled supremacy in the East were numbered.

After the loss of its last trade monopoly, China became mostly a headache for the company, due to a series of lingering unsolved problems. The empire of Great Britain owed its entire acquaintance with tea, not to mention its continuing supply of it, to the empire of China. The Chinese picked tea, roasted it, blended it, and then sold it to England at a lucrative markup. China was in complete control of the drink that had dominated British taste for two centuries. Dependence on another country for a vital product was a blow to imperial Britain's sense of self-sufficiency. It was especially galling to be so reliant on a nation that was so often churlish and disobliging, and that would raise prices on inferior goods whenever it pleased.

To East India House, it had long seemed preposterous that any country, let alone one of "backwards Orientals," could so thoroughly ignore trade initiatives from a nation whose mighty navy, at the time, dominated the world. Yet isolationist China had managed successfully to keep Great Britain at arm's length, even though Britain purchased one out of every five chests of tea it manufactured. Despite a century of diplomatic approaches, the Chinese had yielded absolutely no secrets about the manufacture of tea. How it was grown, by whom, and in what conditions remained a mystery to the West. Even the very names for tea were enigmatic: Lark's Tongue, Dragon's Well, Jade Girl Peak, Looking Glass Rock, Water Tortoise Stones, Rock of Three Monks. Were these green teas or blacks? How could one label them for the market with such descriptions? How could the company be sure the teas would taste the same from year to year? Its Court of Directors had long since grown tired of dealing with grasping middlemen and did not want to share any part of their profits with the infuriating Chinese.

Tea was the symbol of the one major country on earth that still resisted Britain's empire.

If the tea trade had been the greatest boon for the East India Company in its prime, by the mid-nineteenth century the company was reeling; its tea trade was in decline while simultaneously its last, best hope for survival. In India the company presided over a series of human catastrophes: a famine that killed more than ten million and heavy death tolls from warfare between states, corruption, expansionism, drug dealing, and ethnic cleansing. Governing a subcontinent was also an expensive business. To safeguard its territory the company mounted military campaigns in Afghanistan and the Punjab, with scant resources to pay for them. New colonies were supposed to create new markets for British-manufactured goods, but there was little demand from Asian peasantry for British woolens.

The company clerk kept his hand steady as he measured out the chest of Indian tea in the dark-paneled confines of East India House. This was Himalayan tea, which was never before seen in England. It had been sent from Calcutta on the orders of Viscount Hardinge himself as part of a plan to save the company.

Hardinge had battled Napoleon at the sides of both Lord Nelson and the Duke of Wellington. He missed Waterloo by a mere two days after losing his hand at the Battle of Ligny. The viscount was such a favorite of Wellington's that he was nevertheless presented with the gift of Napoleon's sword. Hardinge went from the military to a life in politics, serving as a member of Parliament and later at cabinet level as Secretary for War, in both Whig and Conservative administrations. At the tiller he was an able hand—even if he only had one—and the Honourable Company counted itself lucky to have such a trusted soldier helming India as governor-general from 1844 to 1848. When he suggested that this Indian tea ought to be sent to London's tasters and blenders, the finest practitioners in the entire world, the Court of Directors of the East India Company hastily and heartily agreed.

Himalayan tea was not the first tea produced on Indian soil. The company had been growing tea there for at least ten years, propagated out of native Indian tea plants in Assam Province. Indian tea was initially discovered in Assam by company medical corps surgeons as early as 1815 but was not formally acknowledged as such until 1831. Indian tea grew well in its sea-level home soil near Burma, where the natives chewed rather than drank it. In the following years the company invested millions in the experimental cultivation of native tea, to see if it could be grown in Indian gardens for the domestic market. To an extent the plan worked. The company discovered it could produce a leaf that

looked like the tea leaves of China; it could also train natives to
pick and prepare tea. But the company could never make Assam
tea taste good—or at least not as good as China tea, the finest in
the world and the only one that mattered to a thirsty English
market.

Assam tea had a strong bite to it and a hot, sooty taste. Even
today Assam is seldom graded fine at auction and is appreciated
only by those who desire a strong nose and a certain maltiness to
their brew. It also doesn't grow particularly well, yielding a poor
return per acre. Now as then, Assam tea is largely used in blends
when the prevailing notes of dainty florals require a slight heft.
Within a few years of commencing their tea experiment, the
company realized that Assam would never fetch the high prices
of its rival and would certainly never overtake China tea in the
world marketplace, and so it begrudgingly divested itself of its tea
assets there.

Himalayan tea was the company's next big hope, enthused
Hardinge in a letter to the Court of Directors dated September
20, 1847.

> I consider it highly probable that in the course of a
> few years, the cultivation of [Himalayan] tea is likely
> to prove a highly valuable source of revenue for the
> state. No apparent difficulties exist to the spread of
> Tea cultivation in the Hills to an almost unlimited ex-
> tent and I have every confidence that at no remote pe-
> riod Tea will be produced in sufficient amount not
> only to meet the probably large demand in India but
> also in quantity and sufficient fineness in quality to
> enable it to compete with the Tea of China in Euro-
> pean markets and to render England in some degree

independent of a foreign Country for its supplies of this necessary of life.

The Himalayas possessed the same growing conditions as China's best tea regions. They were subtropical, on roughly the same latitude as Cairo, but high and cool, so the tea would be slow-growing and retain its pungency. There was also unlimited space on Himalayan hillsides for tea production—the natives seemed neither to want nor to make use of the mountains, for food or profit. Under Hardinge's orders, the company made elaborate plans for experimental plantations of a minimum of five hundred acres each, which would allow for economies of scale, capital investment, and the European efficiency that Chinese tea production lacked. British laws and British investors would oversee the sale and merchandising of Himalayan tea; there would be no middlemen, no double-dealing, no Chinese-style obfuscation. Labor in India was at least as cheap as in China, both countries having a surfeit of manpower. The quality would eventually improve, and the prices would drop so that leaves that were picked for a penny could be sold for £3 in London. Growing tea would be like printing money.

As the clerk closed the bags of tea, he sealed them with the wax stamp of the East India Company. Each was sent to one of the esteemed tea brokers of London, the blenders, tasters, and traders whose noses and palates determined the price of a commodity and the fate of nations: Messrs. R. Gibbs & Co., Peek Brothers & Co., Miller & Lowcock, and the revered House of Twinings.

The Court of Directors was "requesting to be favoured with their respective opinion of the quality and value of some specimens of tea grown and manufactured in the District of Kumaon, together with any practical suggestions for its improvement which may occur to them."

At last messengers arrived to pick up the deliveries.

The Court of Directors waited patiently for a response.

When the reports came, they were good—extremely good.

Twinings, Gibbs, Peek, and Miller & Lowcock wrote that the Himalayan tea was as fine in quality as the finest of the China teas. The leaves were perfection: beautiful to look at, picked at the right time, light on the tongue, delicate in the cup, and brewed up a rich liquor with a golden hue. The tea would compete admirably at auction; they would stake their reputation on it.

There were caveats, however. As the experts noted, the Himalayan tea was "lacking in fragrance," that is, it did not have the floral nose of the finest China teas. Some of this was a matter of stock. While the Himalayan teas in the sample had been raised from Chinese seed, it was not the finest seed from the best regions, but ordinary varieties smuggled out of Canton in the south of China, the only place that Englishmen were then allowed. Tea from Canton was known to be of extremely low quality compared to that of other Chinese tea regions.

Beyond the quality of the tea stock itself, none of the Himalayan tea's other noted faults was inherent; instead, some of the tasters' complaints were attributable to poor processing and manufacture. If the Himalayan tea lacked the perfumed notes of China teas, it was because the latter were packed between other materials, such as jasmine, bergamot, lemon, or verbena, to scent the brew. In addition, the Leadenhall Street tea had been poorly prepared for shipment in boxes that were not airtight. The sea air had doubtless tainted the sample, deadening its flavor.

The company's prototype tea may not have had the ultimate refinement, but if the methods and practices of the world's finest

Chinese tea manufacturers could be imported to the plantations in India and if true native Chinese experts could train the Himalayan growers in the processes of tea manufacture, then the deficiencies of Himalayan tea could be profitably redressed.

In 1846 (the season in which Hardinge's tea was grown) the company's experimental tea gardens in the Himalayas totaled little more than six hundred acres, but the court had plans for rapid expansion. The government of India had over one hundred thousand acres ready for cultivation. From such acreage the company could expect to bring in a profit of almost 4 million rupees a year ($100 million today). But to achieve that level of return in only six years—the time it takes a tea plant to mature to picking stage—it would need hundreds of thousands of Chinese seeds from the finest of China's green and black tea regions immediately.

For the Himalayan tea experiment, the company shopping list was short but precise: It sought China's materials, her best seeds, and China's tea knowledge, in the form of Chinese tea makers and tea manufacturing implements.

The company was well aware that getting tea out of China would be a difficult undertaking and impossible to achieve through normal diplomatic channels. As Her Majesty's consul in Shanghai, Rutherford Alcock, warned Viscount Hardinge, "It will suggest itself no doubt to your Excellency that the Chinese are likely to regard any demand on my part for tea seeds or plants with great jealousy, and that the attempts in conjunction with efforts to obtain seeds, to induce Chinese skilled in the cultivation and manufacture of tea to leave their country and proceed to India for the purpose of instructing people, must inevitably fail." In other words, if the East India Company wanted tea for India, it would have to steal it.

Tea met all the definitions of intellectual property: It was a product of high commercial value; it was manufactured using a formula and process unique to China, which China protected fiercely; and it gave China a vast advantage over its competitors.

The notion of intellectual property and trade secrets had been articulated only a few years earlier when a Massachusetts judge ruled in an 1845 patent case that "only in this way can we protect intellectual property, the labors of the mind, productions and interests as much a man's own . . . as the wheat he cultivates or the flocks he rears." In the dawn of 1848, the East India Company was planning a project that was nothing short of industrial espionage. If the company's scheme was successful, the largest multinational corporation in the world, the East India Company, would enact the greatest theft of trade secrets in the history of mankind.

3

Chelsea Physic Garden, May 7, 1848

On a spring day in 1848, Robert Fortune strolled through the Chelsea Physic Garden, a verdant patch of land by the Thames, admiring some of his own handiwork. The earth was just warming to life: Tulips were out in full flower, and lily of the valley dipped gracefully toward the ground. The bulb beds planted in the chilly previous autumn were coming into bloom, as was a tree peony, one of his most treasured discoveries in the Far East.

Three years earlier, upon his triumphal return from China, Fortune had been appointed curator of the Chelsea Physic Garden—something of a vindication of the Royal Horticultural Society's assessment of the value of the mission to his future career.

Now thirty-five, he had seen his life change very much for the better. *Three Years' Wanderings in the Northern Provinces of China* had been published a year earlier to rave reviews. He and his family currently enjoyed an enviable position that had a degree of social prominence and comfort to which he could not otherwise have aspired. His position at the Physic Garden paid £100 per annum (roughly $10,000 in today's dollars), no more than he had

earned in China, but in addition to his salary he received the use
of a charming brick house on the garden's grounds for his family
and servants (though the house had no indoor plumbing or sanitation), an allocation of coal, and the right to cultivate his own
vegetable garden.

The curatorship of the Physic Garden provided him with
both a showcase for his talents and the opportunity to establish
himself as one of the premier horticulturalists in Britain. The
Physic, established in 1673 by the Company of Apothecaries,
was England's second oldest botanic garden. Today it is a bucolic
oasis of only four acres, situated in the posh area near Sloane
Square, not far from the center of London. Hidden from the
street behind high redbrick walls, it serves as a living museum of
the exotic and medicinal plants of the world. In Fortune's day it
was a living display of the many novelties and mysteries of the
Victorian plant world as well as a laboratory for the study of materia medica: herbal and vegetable remedies, balms, powders,
syrups, tinctures, salves, and ointments. The Garden of Simples,
as it was originally known, was instrumental in the development
of horticulture, producing *The Gardener's and Florist's Dictionary,
or a Complete System of Horticulture,* which was the definitive
manual on gardening technique and cultivation for gardeners
around the world for over a century. Alongside Kew Gardens,
just a short journey down the Thames, the Physic Garden played
a major role in the growth of the profitable and strategically important plant-based industries that helped drive the economy of
the British Empire.

The concepts of plant exchange and empire had, in fact,
evolved in tandem. Botanists accompanied Captain James Cook's
first circumnavigation on the *Endeavour* in 1768. Cook went on
to discover Australia while onboard scholars mapped the transit

of the planet Venus, collected samples, and painted pictures of the strange-looking plants of the Southern Hemisphere.

Joseph Banks, a millionaire horticulturalist, was among those who sailed with Cook. As a man of influence, Banks campaigned to put a naturalist aboard all future expeditions of England's world-winning naval fleet. Both Banks and Cook published accounts of their trip; unlike Fortune's, their copious notes did not record their subjective impressions but instead contained precise and descriptive catalogs of their findings. These empire builders brought concrete knowledge from faraway locations, and with the acquisition of that knowledge England gained a growing confidence that it could possess, command, and profit from the entire world.

Plant exchange was a major source of income for the British Empire, which then consisted of old colonies, such as the West Indies, and the recently unified colonies of the Indian subcontinent, along with island outposts in the oceans between. Botanists such as Fortune were charged with suggesting how newly discovered plants in foreign dominions could be exploited for the Western market, how cash crops could be improved through selection and hybridization, where on the globe to cultivate a particular plant to achieve maximum yields from cheap colonial labor, and how to process a plant for market distribution.

Plant hunters were highly trained, sharp-eyed men who left home and family for the lure of discovery. In the opening years of the industrial era, botanic research was a counterpart to today's industrial research laboratories. Botanical imperialism was a way of making colonies pay their way, and plant hunters became the research and development men of the empire.

Although science was very much at the core of Fortune's work, he was at heart a gardener, and a gardener is an artist: His canvas is land; his medium, plants. A gardener works in a three-dimensional world, taking into account the relative heights of trees and depths of borders, the slope of a hillside, and the views to be "borrowed" or enhanced. But he works in a fourth dimension as well: time. A gardener plans for seasons: which trees will bloom in spring (forsythia, magnolia, cherry, lilac, and apple) and which will reach their peak of color in autumn (acer, euonymus, and elder). A gardener's art also spans years—in determining which trees mature quickly and grow tall easily, such as birch, ash, and the softwood evergreens such as cedar, fir, and pine, and which grow slowly and with some effort to leave a lasting legacy, such as oak, beech, and maple, which stand for generations. Fortune was well aware that to be a great gardener demanded great patience.

Gardening appealed to the gentler side of his character. He was a spirited man who enjoyed outdoor living and had an innate sense of what a plant needed to thrive: shade or sun, amount of water, whether to plant on a sloping hillside for drainage or in a container so as to coddle and warm the roots. He could kneel in the soil and know, from years of practice, exactly where to cut back a bush and how to gently bring on a bud. Plants thrived for him.

On that spring day in 1848, Fortune could look upon the reawakening Chelsea Physic Garden with a sense of accomplishment. By his account the place was in disrepair when he took it over—its borders overgrown, its greenhouses in decay, its catalogs worthless. Fortune, with his knack for organization and his own aspirations, brought about a notable transformation. As he wrote to the Garden Committee of the Society of Apothecaries: "From various causes with which the Committee are doubtless acquainted, the Garden has been allowed to get into a most ruinous condition. When I took charge of it . . . I found it overrun with weeds, the Botanical arrangements in confusion, the exotic plants in the Houses in very bad health, and generally in a most unfit state for the purpose for which it was designed."

Fortune cleared the weeds, bought new tools, built up collections through donations and plant swaps with other gardens, and, most important, sold off £364 worth of bank and nursery stock (equivalent today to $44,000) to raise money for the erection of new greenhouses and the repair of the standing ones. His building plans were timely, for the year 1845 had seen the repeal of England's "glass tax," and he seized the opportunity of lower-priced glaziers to order new glass greenhouses to abut the garden's high brick walls. These semi-glazed constructions were soon home to the most exotic new horticultural imports: delicate orchids and ornamentals, spiky bougainvilleas and potted palms, prehistoric ferns and brand-new begonias, balsa, breadfruit, bananas, and bamboos.

The Physic Garden was still arranged in the formal seventeenth-century pattern, with plants heavily pruned and placed in geometrically aligned beds. Fortune wanted to move away from such excessive formality, though, and reorganized the medical garden along the precise and ordered lines of Linnaeus's classification system.

The Victorian age of exploration fed upon an enthusiasm for the natural sciences generated by the work of Swedish naturalist Carl von Linné (1707–1778). Writing under the Latin name Carolus Linnaeus, the scientist invented a taxonomy using two names to categorize the world according to the characteristic sexual organs of plants and animals. Linnaeus's work divided life on earth into kingdoms, kingdoms into phyla, phyla into classes, classes into orders, orders into families, families into genera, and genera into species. A species' name alone provides a great deal of information about it: If an organism is in the *mammalia* class, it has hair and secretes milk; if two species belong to the same order, they are more closely related than a third that is merely in the same class. When a naturalist discovered, say, a new species of beetle on a trip to the Amazon, he would now pose standard questions involving its body type and method of reproduction in order to give his discovery a name. Linnaeus's simple distinctions brought hierarchy and organization to the natural world in a way that had previously eluded scientists.

While pious Linnaeus hoped that establishing the relationships between living things would bring him closer to an understanding of the Creator, his work instead founded a scientific revolution. Linnaeus fueled Europe's burgeoning sense that all things on earth could be comprehended and mastered by the rational efforts of mankind. It was a landmark Enlightenment moment.

Under Fortune's new scheme for the Physic Garden, medicinal plants were displayed relative to one another in their natural orders. "What labour is more severe," wrote Linnaeus, "what science more wearisome, than botany?" Fortune would no doubt have agreed with that sentiment, but as wearisome as the effort often was, it was infinitely rewarding to a man with a mind bent toward order and understanding. One could walk through the Physic Garden and see there a physical manifestation of the march of science and human learning in the relationships between a living organism and its neighbor. To study the beds of the garden was to see natural history codified in bloom.

It is perhaps no wonder that in such a turbulent time as the Industrial Revolution, gardening became a national obsession in Britain. Patience and time were slowly being eroded across the country as technology brought a new immediacy to everyday life. Where a length of cloth, a blanket, or some bedding once took long evenings by firelight to create, countless yards of fabric were now spun each day in the mills of Liverpool and Manchester. Where a trip across counties was once a marathon involving several coaches, it was now a single short ride away by train. Candlelight gave way to gaslight, wind power to steam; the world grew ever more mechanized and reliable. The vicissitudes of weather were becoming things of the past, and as natural processes faded from view, they began to be fetishized. The new middle class of the industrial Victorian age regretted this estrangement from nature and, mourning its loss, was soon willing to pay a premium for a simulacrum of it.

As curator of the Chelsea Physic Garden, Fortune had effectively ascended to the highest position available to him—and he might reasonably have feared that it would be his last. Although he enjoyed the success conferred on him by his China trip, he still faced the limitations imposed on him by his birth and class. A contemporary of Fortune's wrote, "People without independence have no business to meddle with science. It should never be linked with lucre." There were only a few paid positions within the nexus of loosely connected botanic gardens of the empire: Kew, St. Helena, and Calcutta. Although there were some rare and wonderful jobs in botany—working for the Royal Horticultural Society or teaching at University College or running the botanic gardens in Edinburgh and Oxford—on the whole such positions were few and highly competitive. The university appointments went to men with advanced educations, and they were seldom well enough endowed to be a sole means of support. Many of the naturalists who would make names for themselves, such as Charles Darwin, had enormous private incomes with which to fund their studies. Dilettante country pastors and doctors considered themselves naturalists, too, treading the local hillsides to build their collections and often assembling considerable libraries to advance their scientific inquiries. Some of the prestigious jobs in botany were passed from father to son and often to grandson, such as the directorship of the Royal Botanic Garden at Kew, which would remain in the hands of the Hooker family for a continuous sixty-four years.

Whatever the professional obstacles facing him, Fortune at least had the support and companionship of his wife, Jane, née Penny, a lively Scotswoman. He could not have advanced even as far as he had without her help. The curatorship of the Physic Garden brought him a salary, a home, and a vegetable garden, and Jane tended to the latter two. While he was attending to his botanical duties and research, she sowed vegetable seeds in March, moved plants in and out with the sun throughout the spring, and transplanted seedlings after the final frosts in May. She grew the food that her family would eat. She mended old clothes and sewed new ones for a man who all too often found himself in a thicket of thorns, snagging trousers, socks, and coats.

Jane also served as the Fortune family's secretary and accountant. While Fortune was in China collecting plants, his salary was directed to Jane in London. She paid his debts and put money by, managed expedition accounts, and settled his bills. She was also the go-between for Fortune's shipments of trinkets that were sent home to auctioneers. It is entirely likely that she would have kept abreast of the newest developments in botany, too, and forwarded relevant papers and magazines to Fortune's *poste restante* addresses abroad. Away for years at a time, he could not afford to be uninformed of the scientific developments.

The Fortunes had been married for nearly ten years by then, but owing to long periods of separation and to misfortune, they had only two children: John Lindley, named for Fortune's friend and botanical mentor, was four years old; Helen Jane was seven. Helen was already growing into a young lady, her mother's pride and comfort when Fortune was in China, while John Lindley chased after her, pleading for attention. But there had recently been a death in the family: Agnes, named for Fortune's mother, had passed away at less than a year old. She was bright and cheerful, an easy baby to love, and was the first child to be born at the Physic Garden. Her loss was a keen blow to both parents.

If his wife was an essential linchpin in Fortune's life and career, he recorded little about their private life together. Although he has left ample documentation of his time at the Physic Garden—what he built and spent, planted and reaped—no personal memories remain from his time in London. In his published work on China he could present whatever face he chose to the world and fully control the message he delivered. At home in England he would be exposed to close scrutiny, so in his typically taciturn way he gave others as little information as possible by which they could judge him.

It is unclear why Fortune was so silent on his domestic life, given that he portrayed himself as such a lively character abroad. Perhaps the quotidian details of his life seemed too confined compared with his adventures in China. The Middle Kingdom at the Center of the World, the greatest research laboratory he had ever known, still called to him. It is also possible that his habitual secrecy arose from his sense of shame about his past. It was not just that Fortune's beginnings were humble but also that he could not rise above the common but telling discrepancy in the parish records of his birth. Robert Fortune was born on September 16, 1812, but his parents, Thomas and Agnes (née Ridpath), had only married on June 24, 1812. The condition of his mother, seven months pregnant, would hardly have gone unnoticed as she walked to the altar in a small town in rural Scotland. As Fortune rose to prominence as a public figure, the details of his date of birth would be altered—from 1812 to 1813—perhaps to preserve a semblance of propriety.

It was on May 7, 1848, that Dr. John Forbes Royle paid a momentous visit to the Physic Garden. By then an old man, Royle was among the most esteemed figures in botany and was professor of materia medica at Kings College, London. He was closely acquainted with Lindley, Fortune's mentor, and a member of the prestigious Royal Society and the Linnaean Society. Although he had hosted eminent guests in the past, Fortune was nonetheless pleased to receive a man such as Royle and the compliment his appearance implied.

Royle had come to see Fortune on behalf of the East India Company, as their horticultural adviser, to discuss the subject of tea. Also a Scot, Royle had grown up with the company and was practically raised in it, having attended its military academy at Addiscombe. He went to India in 1819 and, upon discovering the joys of botany, declined his military commission to become a surgeon, being eventually placed in charge of the botanic gardens at Saharanpur in northern India. Royle's recommendations in *Illustrations of the Botany and Other Branches of Natural History of the Himalayan Mountains* and *An Essay on the Productive Resources of India* led the company to establish an entire department devoted to botanical concerns. Royle's knowledge of the growing capaci-

ties of the Himalayan range was unmatched by any other botanist. He believed, along with Hardinge, that tea could very
profitably be grown there.

Royle and Fortune walked together to the far wall of the garden to examine one of the new greenhouses. Royle, who had not
been to the subcontinent for many years, admired the glass
gleaming in the sunlight. These warm environs hinted at a skill
that the East India Company required. Fortune, on his trip to
China, had become an early expert in a new technology. As previously mentioned, it was called the Wardian or Ward's case and
is known today as a terrarium. Portable glass houses such as this
would change the growing patterns of the planet.

Prior to about 1840, plants were poor travelers, and plant exchange between the colonies of Great Britain was difficult and
often impossible. Seeds and live cuttings from abroad spent
months on a ship, crossing the equator at least once and often
twice on their way back to England or elsewhere. Sailors were
not trained as gardeners, and plants consequently did not travel
well under their supervision. Fresh water was scarce on a long sea
voyage and was not easily surrendered to exotic flora. Often
stowed on deck in direct sunlight, the plants were also exposed to
corrosive sea spray. If they were stowed belowdecks, away from
the sun, they died the slow death of deprivation. It was a rare and
hardy specimen that could survive an ocean voyage.

But Dr. Nathaniel Bagshaw Ward of London had changed all
that with a series of papers that attracted the attention of professional naturalists. In the late 1830s, Ward had made a catalyzing
discovery: In a closed bottle where he kept a hawk moth chrysa-
lis, he saw seeds germinating on a piece of common mold. The
seeds had been sealed in the bottle the summer before, were kept
warm and protected, and had not been touched since then. With-

out opening it Ward transferred the bottle to a window ledge and took careful note of further developments. Within four years the seeds had sprouted into a fern and some common grass.

Born in 1791, Ward was the son of a doctor and was raised in the docklands of London. The sailors who frequented his father's practice must have awoken in the boy a taste for the faraway and exotic. At his own request Ward set sail for Jamaica at the age of thirteen, and although the harsh life of a sailor quickly palled for him, the tropics continued to fascinate him. Like so many other travelers, he would become a botanist, an herbalist, and a man of medicine.

Ward was also an obsessive. For years after the chance discovery of the seeds in the bottle, he experimented with glass, seeds, and mold, taking careful notes of his observations. Whatever plant he chose to raise in a glass container thrived. (The songbirds he included in these experiments did not, however.) His herbarium grew to include twenty-five thousand specimens, and still he kept testing. Ward was the first to recognize a fact that had previously been unimagined: Plants can survive for years kept in a sealed, well-lit environment without water. He had a series of glass boxes made, kept airtight with putty and paint. He was stumbling upon the missing piece of a vexing puzzle: how to keep plants alive during long, arduous ocean transits.

Ward witnessed and documented a process that was simple and self-sustaining: During sunlight hours, plants use moisture from the soil in combination with carbon dioxide to photosynthesize. At night they emit oxygen and release water vapor, which condenses in the cool night air against the glass and drips back down to moisten the soil. The moisture is almost indefinitely retained, so plant life in such containers is effectively self-perpetuating.

For professional plant hunters the implications of Ward's discovery were revolutionary. Previously, there were few ways to show anyone a live foreign plant—no good, reliable method for taking it any great distance for study existed. A naturalist could examine plants by killing them, by digging up a specimen and drying or pressing it. Or he could attempt an artistic rendering. To preserve live specimens, the traveling botanist's job was one continuous gamble. Which seeds would stay fresh and which become waterlogged in wet sea air? Which seedlings were robust enough to transit the tropics as well as the northern zones? Before the Wardian case, the foreign plants that grew in Britain were those few hardy varieties whose seeds and seedlings could withstand extremes of temperature.

Now for the first time in history, naturalists would be able to preserve plant life ex situ.

In 1834, when Ward was in the middle of his experiments, a ship docked in England on a return trip from Hobart, Tasmania, with several of the new glass cases on board. Despite the range of climates the ship had traversed—several winters and summers in one crossing and many thousands of miles, with sea spray and salt air buffeting the boxes the entire way—the plants arrived intact. Such was the excitement engendered by Ward's discovery that ships bearing his cases were soon launched to the far corners of the world.

"To sum up all," Ward wrote of his experiments, "in every place there is light, even in the centre of the most crowded and smoky cities, plants of almost every family may be grown."

Ward's discovery had enormous economic ramifications for the empire. For instance, the bark of the Peruvian chinchona produces the alkaline quinine. This tree could now be transplanted to the subcontinent, and locally produced quinine could be used to treat the malaria that plagued the British soldiers of India and Burma. Brazilian rubber trees, raised by seed at Kew Gardens, could be replanted on the hospitable island of Ceylon (now Sri Lanka) and generate a new source of revenue. Entire industries could germinate in Ward's glass cases. Even hobby gardens were taking on a different look as hardy small specimens of trees from abroad, such as the cherry and the flowering crab apple, were liberating the gardener from the tyranny of lavish annual bedding schemes grown from seed. The fashionable horticultural enthusiasts of Britain, including Prince Albert, eagerly prepared for an influx of millions of new plants. And the East India Company was ready to wager that the Wardian case could help them transport the finest Chinese tea plants and seeds to India.

Royle and Fortune spent a long afternoon in the garden talking about the Wardian case, the Physic Garden, China, and Fortune's brilliant book. China held a special place in the imagination of plant hunters, for it was an entire nation of gardeners. Unlike much of the world undergoing Britain's program of colonization, China was, to the British mind, almost civilized. The Chinese had cultivated passions and refinement, poetry, music, and philosophy; above all China had a reverence for gardens. Mandarins displayed their status by building winding gardens among fishponds, stone bridges, and pavilions in which to meditate on Confucius. Chinese peasants knew how to grow their own food and how to forage for wild edibles. China's geography also promised a bonanza for plant hunters. An enormous country with a variety of hardiness zones, from temperate to tropics to tundra, and vast changes in topography, it was an unparalleled natural showcase for genetic variation and natural selection.

The grandfather of British botany, Sir Joseph Banks, had viewed China as the Holy Grail of plant hunting. Banks had arranged for a gardener to attend Britain's very first diplomatic delegation to the Chinese emperor in Peking in the late eighteenth century. A gardener could "never fail of learning something, if he can be brought into contact with his brethren in Pekin," Banks wrote. He sent a blanket appeal to all Englishmen in China, amateurs and experts, diplomats and sailors, for plants that were "either useful, curious, or beautiful" and requested they bring these home in any way possible.

Banks's notion of the botanical bounty of China was largely based on rumor and supposition, given that there was so little information available from anywhere but the southernmost port of Canton. He nevertheless asked for specific details regarding the Chinese method of dwarfing trees—or bonsai, as we now know it as practiced in Japan. Among the plants that Banks hoped would be collected were various azaleas, the Moutan tree peony, the lychee, the longan nut, economically valuable plants such as tea bushes, and hardwoods such as oaks. Britons in China were also asked to investigate Chinese methods of turning human waste, or "night soil," into fortifying garden manure. England, with its rapidly growing population and lack of a working sewerage system, had a surfeit of human excrement. The introduction of this particular Chinese technology could help turn a public health nightmare into a productive boon for industrial Britain.

"To leave behind a once scarce and curious plant under the mistaken idea of its being a common one will be a source of vexation forever afterwards if the circumstance happens to be discovered," Banks threatened.

On his own China trip Fortune had effectively fulfilled Banks's directives nearly sixty years after they were issued. But had he left important plants behind? Although he journeyed widely in his first three years, visiting places no Briton had ever been, he traveled mainly from treaty port to treaty port. He had done at least as much collecting in the markets of Chinese cities as he had in the fields. Fortune and Royle marveled at the fact that the Chinese interior, even after Fortune's assiduous collecting, was still essentially untouched and ripe for exploration.

Now Royle had a proposition to make: Would Fortune be willing to return to China in the employ of the East India Company as a tea hunter?

The company's terms would be very generous. In contrast to his current salary of £100 a year, what a starting clerk was paid in the city, Fortune would receive £500 per annum (about $55,000 today), which was equivalent to the wages of a man who had worked in a trusted position for twenty-five years. His passages out and home would be paid, as well as all other travel expenses—including the cost of cargo shipped between China and London. Cargo space was precious to the plant hunter; although the goods he would carry home amounted to little more than market produce, transporting them was the single greatest expense incurred in any botanical exploration. Every novel and curious plant had to compete with the profitable teas and silks also vying for the limited space available to hire, and the prices on stowage were raised accordingly.

But the most magnanimous term of the company's offer was simply that Fortune's remit was so narrow: He was being engaged only to collect tea. The property rights to all the other plants he collected—the ornamentals, grasses, seeds, seedlings, flowers, fruits, ferns, and bulbs—would be Fortune's alone. With such generous conditions it would be possible for him to start collecting on his own account and to sell samples at auction for potentially vast profits.

As the gardening obsession of Great Britain grew, a softer,

romanticized "English landscape" became fashionable. In country estates, vistas were plotted, lakes were dug, and hills were arranged. These new artificial tableaux demanded rare specimen plants to enhance them. While Fortune regarded such gardening fashions as trivial, he cannily recognized the potential of the company's proposal for his own personal advancement. The gardens of Britain were changing: Auction rooms were filling up with plants from abroad. Competitive amateurs and experts were bidding up prices on the exotics brought back by plant prospectors. If Fortune could go back to China as a collector and actually sell his discoveries, as so many of his contemporaries in other parts of the empire were doing, he could become a rich man.

He would give it due consideration, he told Royle.

As the eminent doctor left, Fortune stood at the wall near the iron gate, looking at the climbing pale pink roses, budding but not quite in bloom. The English rose of legend had originated in Persia and had come to the British Isles only a few hundred years before, but China had been cultivating roses for centuries. Only fifty years earlier an accidental cross-pollination between the two varieties produced what is known today as our common garden rose: long flowering, sweetly scented, hardy, and low growing. It is said there were no deeply colored roses in England before the introduction of the China rose; the War of the Roses could have had no true red rose for a symbol, only a pale pink one. The pedigree of England's roses was only one example among many in Fortune's Physic Garden of how flowers from the East had hybridized and changed the plants of the West.

Elsewhere in the garden Asian favorites flourished: The lilac came from Persia, the tulip from Turkey, and citrus from Southeast Asia. And thanks to Fortune himself, flowering camellias had come to England from China.

But what the company had now requested of him was a much bigger undertaking than simply collecting. Fortune would have to steal samples of one of the world's most economically valuable plants, keep them healthy, and arrange for their successful trans-plantation on another continent. It was the most formidable task a botanist had ever faced.

He would have to speak to Jane.

Barely one week later a letter arrived from East India House.

To Mr Robert Fortune
Botanic Gardens Chelsea

Sir,

With reference to the communications with Dr Royle on the part of the Court of Directors of the East India Company has held with you upon the sub-ject of your proceeding to China for the purpose of obtaining plants and seeds of the best descriptions of tea from the most desirable localities and of conveying them under your own charge from thence to Calcutta and eventually to the Himalayas . . . I am commanded by the Court to acquaint you that they accept the offer of your services and that they will expect you to be ready to proceed to China not later than the 20th of June next.

The Court will grant you a salary of Five Hundred Pounds per annum to commence from the date of your embarkation and to cease on your return to this coun-try. They will provide you with a free passage to China and you will be entitled to a free passage on your re-turn to England. The court will also defray all your travelling charges and other expenses which you may incur in India and China in procuring and conveying plants and seeds and in otherwise carrying out the objects contemplated by the Court in view to extend the cultivation of tea in the hill tracts of the North West Provinces of India.

As it is of importance that you should arrive in China as early in the Autumn as possible a passage will be procured for you in one of the Peninsular and Oriental Company's vessels in order that you may proceed in the most expeditious manner.

East India House,
17th May 1848

4

Shanghai to Hangzhou, September 1848

A flat-bottomed boat was moored in a snaking, stinking canal one day's sail out of Shanghai. The boat was small, no more than 40 feet long, a floating home belonging to a seagoing family of brothers and their wives. The family shared the labor and meager profits of conveying cargo and travelers, illicit or otherwise, through China's coastal network of waterways and canals. It was a creaky junk, unremarkable in the environs of Shanghai except for its current group of passengers.

Fortune's Chinese body servant, "a large-boned clumsy fellow," sat threading a blunt needle with horsehair. He slipped the needle beneath the hairs at the nape of Fortune's head, yanking the stitches taut with every pull. He was sewing a long braid, black and coarse, formerly the pride of some peasant, into Fortune's own hair. The queue, as it was called, hung from his neck to his waist as if it were his own.

Fortune sat stiffly but nervously, for although he had been in China for several weeks, his true journey was only just beginning. His progress toward the little boat had been efficient and swift: After speeding through Hong Kong—which on his last trip he had dismissed as a "barren island, with only a few huts upon it," and now described as a remarkable British outpost of "palaces, and gardens too"—he journeyed by steamer to Shanghai, a city bustling with newfound foreign trade. In the British concession area there, on the banks of Shanghai's Huangpu River, Fortune took up a brief but productive residence in the palatial home of Thomas Beale, of the renowned trading company of Messrs. Dent, Beale & Co., an old friend and intellectual benefactor to many British travelers in the East. Beale gave Fortune a free hand with his firm's compradors, the trusted Chinese fixers who in the ports regularly matched the needs of Western merchants to the capacities of China. As the boomtown city opened to Western investment and settlement, these translators and profiteers served as envoys who could smooth relations with China's business elite. While in Shanghai, Fortune hired two servants, collected provisions, ordered and directed the assembly of glazed Wardian cases, and planned for a trip that was in reality beyond the scope of precise planning. With little useful reliable information on the Chinese interior, Fortune hoped to determine where the best tea was grown and how to get there with the assistance of those who were familiar with the areas, and he had found it in his choice of servants.

These servants were his entrée into China; they would share the task of collecting the tea as well as, he hoped, of keeping him alive. He had hired two men from the most celebrated green tea districts in the vicinity of China's famed Yellow Mountain. They would act as his interpreters, cooks, botanical collectors, bodyguards, porters, and, most important, guides. The men divided up the traveling jobs roughly between those requiring intellectual skill and those requiring brute force. Fortune's hairdresser, "the coolie," whom his written account never mentions by any other name, was the human ox paid to move luggage from shore to boat and back and carry the heavy glazed cases on botanical collecting trips.

Wang, an educated man in his early twenties, was the more refined of the pair. He was raised on a tea hill near China's finest green tea region, Sung Lo Mountain in Anhui Province. For generations Wang's family had grown and picked tea, sending favored sons off to the cities of Hangzhou and Shanghai to make their name in trade. China's population had doubled in the previous century, and there was too little farmland to support large families. Wang, like many provincial emigrants to Shanghai, was a natural middleman, crafty, seeking a cut from every transaction, and making a living in the gray economy of the newly opened foreign concessions. For Fortune, Wang would prove his value as a professional trip manager. He knew all the roads between Shanghai and the tea districts. As a born businessman he negotiated contracts with porters and boatmen. And as was common in China, he managed to keep a fraction of each transaction for himself, a practice known to Englishmen in China as the "squeeze."

Of the two men, Wang had the advantage in Fortune's eyes of being able to make himself easily understood, which the coolie could not. Wang spoke a language used in the ports that had been developed in the hundred or so years of the tea trade; it was a pidgin composed of English and Chinese, with a smattering of Hindi and Portuguese. Pidgin was at once comical and necessary to the white men who traveled and did business in China. While some words have made their way into the common English lexicon—for instance, "chop-chop" (quickly)—entire sentences often sounded as ridiculous as a nursery language: "Long time my no have see you." "What thing wantchee?" With communication proving so difficult, it is little wonder that foreigners and the

Chinese held each other in such low esteem. Wang spoke pidgin, which Fortune understood, but the coolie was in effect mute to his master because the low-born servant had not learned the basic foreign tongue. In addition, Fortune's own middling Chinese, which he had learned from wealthy men in port cities, was almost incomprehensible to him.

On this mission for the East India Company, Fortune was going deeper into the country than any Briton had ever dared, beyond the reach of British influence. While in the foreign concessions of Shanghai, he operated effectively under the protective umbrella of British law, a quirk of the treaty that had ended the First Opium War. All Europeans inside the trade zone enjoyed special status; they obeyed the laws of their home countries, not those of China. The rule of the emperor could not touch them there.

On this journey far outside the treaty port, Fortune needed body servants more than he had during his first trip, because China was now a more dangerous place. The increased European presence on the coasts was fiercely resented by the local Chinese, and in the south angry peasants had begun attacking foreigners, holding them hostage in factories and hospitals, and sometimes killing them without reason. Rebels were taking over the countryside. The weakening court in Peking, humiliated after its defeat in the foreign war, had lost its control over local mandarins who tyrannized cities and villages with excessive taxes that rose faster than any peasant could conceivably pay.

As the West moved into Shanghai, the Chinese moved out—taking everyone and everything with them, even their ancestors, as Fortune noted: "Their chief care was to remove, with their other effects, the bodies of their deceased friends which are commonly interred on private property near their houses. Hence it was not uncommon to see several coffins being borne by coolies or friends westward. In many instances when the coffins were uncovered they were found totally decayed, and it was impossible to remove them. When this was the case, a Chinese might be seen holding a book in his hand, which contained a list of the bones, and directing others in their search after these last remnants of mortality."

On the face of it, it seemed an entirely foolish venture for a lone European to travel to the interior of China—a fact that was not lost on Fortune's servants. When Wang tried to negotiate Fortune's passage with the junk captains in Shanghai, the sailors refused. Boatmen were regularly beaten and tortured for trafficking the waterways with illegal cargo. "On this account it was impossible to engage a boat as a foreigner and I desired my servant to hire it in his own name, and merely state that two other persons were to accompany him." It was a shrewd plan, and Wang returned with a contract officially signed, stamped with a "chop," or character-bearing seal. But as the traveling crew loaded the ship, the coolie, either from ignorance or malice, revealed to the captain Fortune's identity. Fortune feared the boatmen would no longer consent to having him on board, especially after they had been tricked, but Wang assured his master that the trip could proceed as planned "if only you will consent to add a trifle more to the fare."

Fortune was patient as the coolie attended to his new coif. A small blue and white tea bowl sat nearby on a dusty crate, and swirling its sediment of leaves, Fortune spilled the cooling liquid out onto the dirty deck. Floors were the place to toss garbage in China, it seemed, and he was consciously trying to behave in the Chinese manner to make his disguise credible. And so, in the Chinese way, he had warmed the porcelain bowl by rinsing it with the hot water. Green tea was not Fortune's preference, absent the civilized comforts of milk and sugar, but he was coming to appreciate the custom of drinking it plain and unadulterated.

Fortune was a constant curiosity when traveling as a Westerner. To the Chinese, the Scotsman looked grotesque. He was tall, his nose was much longer than a nose need be, and his eyes were too round; although round eyes were generally considered a sign of intelligence, Fortune, with his halting Chinese, would have sounded like a child to them. Even the simple act of eating brought him unwanted attention. "He eats and drinks like ourselves," observed one member of a crowd, watching him on his first trip, Fortune recalled. "'Look,' said two or three behind me who had been examining the back part of my head rather attentively, 'look here, the stranger has no tail'; then the whole crowd, women and children included, had to come round to me to see if it was really a fact that I had no tail."

Not surprisingly, his servants insisted that they would join him only if he took steps to disguise himself upon leaving Shanghai. "They were quite willing to accompany me, only stipulating that I should discard my English costume and adopt the dress of the country. I knew this was indispensable if I wished to accomplish the object in view and readily acceded to the terms."

The style of the day required that all ethnic Chinese men shave the front of their heads as an act of fealty to the emperor. The tonsure of nearly 200 million people was a potent symbol of the invading Manchurian court's power over the individual. The Qing emperors used the edict as a way of controlling the population, of transforming a multiethnic, heterodox society into a conforming one. Refusing to be shaven was considered an act of sedition.

Having finished attaching the queue, the coolie now took his rusty razor to the front of Fortune's head and began to create a new, higher hairline for him. "He did not shave, he actually scraped my poor head until the tears came running down my cheeks and I cried out with pain," Fortune wrote. "I suppose I must be the first person upon whom he had ever operated, and I am charitable enough to wish most sincerely I may be the last."

On that first day of his journey, Fortune reviewed the itinerary and the rationale for his offensive. The job would require several years in China to complete. To jump-start production in the East India Company's tea gardens, it was critical that he bring back several thousand tea plants, many thousand more seeds, plus the highly specialized techniques of Chinese tea growing and manufacturing. He would somehow have to persuade workers from the finest factories to leave their homes and accompany him to India.

Before Fortune could identify the perfect tea recipe, however, he needed to obtain the basic ingredients themselves: the finest classes of tea that China had to offer, both green and black. To this end he decided to make at least two separate tea-hunting trips—one each for green tea and black tea, for the two were never grown together in the same region. Green tea and black tea required different growing conditions, Fortune believed. The best green tea was in the north, whereas the best black came from mountains in southern China.

Fortune chose to make the first of his trips to the green tea districts of Zhejiang and Anhui provinces. His second would not take place for at least another season, when he would go to the fabled black tea districts of the Bohea or Wuyi Mountains in Fujian Province, traveling as far as two hundred miles inland.

While black tea was the bigger prize for Fortune and the East India Company, given its popularity in the West, it was also more difficult to obtain. It was grown high among the fingerlike mountain karsts, where the thin air and chilly nights produced the richest oolongs, pekoes, and souchongs, the finest black teas in the world. It was at least a three-month trek south from Shanghai to the border areas between Fujian and Jiangxi provinces, a trip that did include the option of remaining out of public view on a riverboat while in transit. No foreigner, save the occasional French missionary, had been to the remote area since the days of Marco Polo.

The logistics of obtaining green tea were relatively uncomplicated by comparison. The districts producing the finest greens were easy to reach, requiring only a few weeks' sail on the great Yangtze River and its tributaries. He had made similar if shorter ventures on his previous trip, traveling by boat until he reached a distant hillside and then wandering with Wardian cases until he had harvested his seeds and dug his fill of specimens.

On his first trip to China, Fortune had learned more about tea cultivation than any other Westerner. He had visited accessible green tea gardens near the treaty port of Ningbo in the company of the British consul. Two entire chapters of *Three Years' Wanderings* are devoted to what he observed there about the growing, harvesting, and processing of the tea plant. Fortune had even brought back tea plant specimens to England's botanical gardens, greenhouse shrubs that proved to be useful for study but worthless for providing any specific information about how tea came to be made. Fortune blamed the "jealousy of the Chinese Government" for his lack of knowledge, because the emperor "prevented foreigners from visiting any of the districts where tea is cultivated." Chinese tea merchants themselves had proved an unreliable source because they were too far down the chain of supply to be of any use to a scientist. His living tea plants in English hothouses could not even settle the ongoing debate as to whether green tea and black were actually different species. "We find our English authors contradicting each other, some asserting that the black and green teas are produced by the same variety and that the difference in colour is the result of a different mode of preparation, while others say that black teas are produced from the plant called by botanists *Thea bohea*, and the green from *Thea viridis*," Fortune had written. A second trip to China would enable him to finally settle the debate.

Fortune hoped to be the first man to successfully plant an entire garden of tea—an entire new industry—from foreign stock. To seed India with inferior tea was hardly going to make him a hero. There was no point in sending to India a motley assortment of inferior tea plants, as previous collectors had done. Fortune was only interested in procuring the most celebrated teas in China: "It was a matter of great importance to procure them from those districts in China where the best teas were produced."

Fortune needed to be as precise as he could about what he had collected and shipped to India. Because he was a scientist, his work was only as good as his data, so he knew he had to verify firsthand his samples' location, ecology, and cultivation. While it occurred to him that he might manage things much more expeditiously and with less risk if he sent local operatives to do the gathering and reporting—as the company had done for generations—Fortune dismissed this option, having little confidence that Chinese agents would be dependable enough to seek the best that China had to offer. And if he did not collect them himself, well, where was the adventure in that, let alone the science?

But it was not sufficient merely to send a shipment of tea plants to India; he would also have to ascertain that the tea he sent not only arrived safely but was transplanted successfully. Three years of labor in China would be for naught if his plants did not survive the journey and thrive in their new locale. To ensure that he received word of the plants' condition as quickly as possible, he had to enlist the cooperation of the government of the North-West Provinces in India, to which he wrote:

> Having been appointed by the Honourable Court of Directors of the East India Company to proceed to China for the purpose of transmitting plants and seeds of the best variety of tea . . .
>
> It is my intention to send down to Calcutta both seeds and plants by variety of opportunities and it will be of great importance if they are carefully received and forwarded to their destination. [It will] also be very desirable to have a report made upon the condition of the plants and seeds when they arrive in India which report could be sent to me for my guidance with regard to the number which it will be necessary to collect.
>
> I trust I will be excused in making these suggestions as the transmission of plants is always attended with some difficulty. . . .
>
> I shall be glad to receive any instructions which you may think it necessary to give me. These [can] be addressed to the care of Messrs Dent who will forward them to me.
>
> I have the honour to be yours . . .
> Robert Fortune

"Hai-yah—very bad, very bad," the coolie muttered in pidgin, razor held aloft in his hand while Fortune winced in pain. "Very bad" and "very good" were the only English words the coolie understood. He took a hot towel and dabbed at Fortune's bloody scalp. The other boatmen sat near the barber's chair, laughing and gambling, the clack of mah-jongg tiles punctuating their conversation. "The poor Coolie was really doing the best he could," Fortune noted magnanimously.

His second servant, Wang, had procured for Fortune appropriate clothing: a gray silk garment that buttoned down the front, with a high stand-up collar; flowing trousers with legs so wide that two men might have walked in them; sleeves that hid his large gardener's hands; and thin slippers, which hardly seemed as if they would stay on a man's feet, let alone protect him from the muck of a city street. Over this he wore a padded coat that was sashed and had deep pouch pockets; it would prove invaluable once he started collecting specimens. It was, all in all, an outfit that any dignified traveling Chinese merchant might wear, inviting respect without attracting undue attention.

Fortune had neglected, however, to request an accounting of precisely what had been spent on these items. Chinese copper money amounted to fractions of pennies, and so Wang had casually pocketed the change unchallenged. The coolie, noting an imbalance in the accounts, was furious, believing that jobs such as handling money—and skimming it off the top—should rightly go to him, the senior of the two, rather than the upstart Wang.

The coolie accordingly tried to complain to his master, who merely laughed off his disgruntlement. His servants' competition for favors made him feel safer; he hoped their bickering would provide a check on any notions they might have of forging an alliance against him. Fortune had assumed that they would be docile and willing servants, and he had given very little thought to them as individual human beings. He had not taken into consideration, for example, that the coolie would endlessly seek esteem or that Wang's impulse to "squeeze" might grow insatiable throughout the course of the journey. Wang was already playing the game of translation arbitrage: All prices were negotiated on paper, written in Chinese numerals, which Fortune did not read.

With his barbering completed and outfitted in his new wardrobe, Fortune naturally wondered: *Would the disguise work?* He was confident that China was so enormous and far removed from the rest of the world that the peasants among whom he would be traveling could scarcely make sense of the sheer size of their own country. With no external reference points for comparison, it would be unlikely that they could determine just how foreign Fortune actually was. He hoped the fact that his facial features were not Chinese or that he was nearly a foot taller than every man around him would not be considered too suspicious in a nation already ruled by foreigners. His height could in any case be explained away: It was well known that the people from the other side of the Great Wall were very tall and extremely brutish, and since all Chinese subjects had queues, Fortune's wearing one would mark him as a subject of the emperor.

The light turned gray around them as the sun sank from view. From shadowed alley doorways a handful of prostitutes began moving toward the canal to set money on fire. It was hell money, burned as a way of transferring it from the physical world to the spirit one; it was offered as an appeasement to deities and ancestors before getting down to the night's business. The coolie cast a wistful eye in their direction.

Fortune felt the braid bounce off his back, dancing lightly between his shoulder blades. He was "a pretty fair Chinaman," he thought, and he reminded himself that from this moment forward he would have to speak only in Chinese, unpracticed as it was after his three-year absence. He would have to use chopsticks and remember to kowtow rather than shake the hand of a man of higher rank. He would introduce himself with his Chinese name, Sing Wa, Bright Flower.

After packing his razor and scissors, the coolie nodded to his foreign master and walked aft to the Sea Goddess's altar, where he lit an incense stick to ask the deity's blessing on their journey.

Wang, visibly nervous as the boat pulled out into the canal and the journey began in earnest, asked Fortune what would happen if someone asked where they were from. How were they to answer?

Fortune smiled and replied, *Wo hui gaosu tamen, wo shi wai-shengren cong changcheng geng yuan de yi xie shengdi lai de.* (I am Chinese, from a distant province beyond the Great Wall.)

5

Zhejiang Province near Hangzhou, October 1848

For the past millennium the Chinese have loved the city of Hangzhou, which stands about 120 miles inland from Shanghai. Built around a lake, with green mountains rimming its skyline and a mist working broad strokes across the water, it is a city for poets. Also a center of serenity, it is liberally endowed with temples and gardens. More than simply a picturesque site, however, it is one of the precious few beautiful places remaining in modern China. The surrounding province was then and remains today the wealthiest in the country. In 1848 its people, from house servants to merchants, were dressed in lush, brightly colored silk robes as they thronged the busy streets, feasting on cakes or trading luxuries such as pearls and jade. Its shops were full and its merchants were fat, which was no mean feat in a nation that suffered from repeated famines and shortages. Where most Chinese cities were filthy and overrun with vermin, by Fortune's report Hangzhou was notably well tended. "Beyond dispute the finest and the noblest [city] in the world," Marco Polo wrote upon laying eyes on it. The Confucian poets were equally enthralled: "Above, there is heaven, but on earth there is Hangzhou."

More than anything else Hangzhou was a city whose life centered on tea: trading it or sampling the choice blends served in the teahouses. It was exactly the kind of atmosphere Fortune would have enjoyed: an ancient gardened place where rich people had the time to not work but to think, to plan big ideas, to enjoy a long cup of hot tea, and to talk about the magic of nature. Were Fortune merely a tourist, he might have lingered here, but he planned to avoid Hangzhou entirely because it was too dangerous to pass through it. A bottleneck for trade, it was situated directly on the main silk and tea routes where merchants from the coastal trading ports might easily recognize a white man in disguise. Or at the very least they would notice Fortune as someone who was an anomaly among the local Chinese.

He therefore commanded Wang to arrange for sedan chairs to bear them around Hangzhou. Once they were safely beyond the outskirts of the city, he was to book them passage on a cargo boat headed up the great Yangtze valley, toward Wang's own home in rural Anhui Province, where green tea grew on every sloped hill.

Wang, however, had other plans: The quickest route to the tea country was to go directly through Hangzhou. The difference in the price of Fortune's passage, the margin between "around" and "through," was more than enough to provide some illicit tea and tobacco for Wang and perhaps even a "flower girl" or two. A little extra squeeze off the top of Fortune's travel budget would buy the coolie's silence.

Wang, as ordered, negotiated with sedan-chair porters to bear Fortune along the long flat road. The "chair" was little more than a box resting between two bamboo poles, held at either end by coolies. "No one took the slightest notice of me, a circumstance which gave me a good deal of confidence," Fortune said of his journey toward Hangzhou.

The road leading to the city stretched ahead for miles, surrounded by forests of mulberry trees, the workshops in which China's famed silkworms spun their soft magic. "I saw little else than mulberry trees," he wrote. He sat comfortably in his sedan chair and, as the long day of travel drew on, was "expecting at any moment to get out into the open country." But with every mile the porters advanced, crossroads became more frequent, buildings sprang up, and fields disappeared. "I was greatly surprised to find that I was getting more and more into a dense town." The porters soon carried Fortune through the gates set in Hangzhou's thick gray walls and straight into the center of the city, next to the splendid West Lake.

"Had it been known that a foreigner was in the very heart of the city of Hang-chow-foo, a mob would soon have collected and the consequences might have been serious," a terrified Fortune later wrote. He immediately upbraided Wang, using the insults of a low sailor, the only form of Chinese swearing he was likely to have known. He was unaccustomed to being disobeyed and probably feared that if he did not come down on his men sternly now, he would face a series of such misdeeds. With every step farther inland, the consequences might be ever more grave.

But Fortune's recriminations had little effect, for a master's scolding his servants publicly only served to build their self-esteem, or "face," as even a reproach was a tacit declaration that the servant was important enough to merit the notice of a wealthy mandarin. In China, "face," or *mianxi*, was a concept that a Westerner like Fortune did not instinctively understand, describing as it does the prestige and reputation one gains from every human interaction. Relationships in China were defined by the reciprocal obligations between people, whether of the same or a different status, and every individual existed within a network of influence, a matrix of duties and social connections, or *guanxi*.

The family came first, then the extended social neighborhood. "Face" expressed a person's position within his or her network and was the mechanism by which the Chinese assessed their obligations: which orders to obey, which favors to grant, and which supplications and apologies to make. A son might perform humble acts for his father, or an employee might bow before his master or a student before his teacher, but in turn the father would have a set of defined responsibilities to the child, the master to the slave, and the teacher to the student.

However subtly they were expressed, *mianxi* and *guanxi* were inescapable facts of life in China; then as now they forged the social fabric of the nation. Social connections determined the measure of justice received and discrimination suffered. While no Chinese person was free from these relationships, many peasants had very little face and therefore little access to justice, wealth, or freedom. When social obligations were met, someone gained face and an increase in status; when a person failed those to whom he was socially connected and thereby obligated, he suffered a loss of face (*diumian*) and a downturn in his social standing. When Wang was shouted at by Fortune for failing to follow orders, it demonstrated to the world that he had responsibilities to an important man. Wang lost face with Fortune, while simultaneously gaining it in the wider neighborhood of Hangzhou.

Face was a very Confucian concept. The great philosopher, whose ideas gained influence during the Han dynasty, 206 BC–AD 220, described a world where familial connections and obligations to ancestors were the highest good and the greatest aim of an individual. A single person was nothing if he did not bring honor to the world from which he came.

A foreigner in China had no network of relationships of prescribed duties and no social capital, and therefore lacked any obvious signifiers of face. Many foreigners handled their outsider status adroitly. They engaged in relationships with the Chinese immediately, offering gifts and favors to officials and higher-ups; they recognized that a servant did not just serve but was owed things other than monetary reward, such as honor and respect. Fortune, however, seems to have paid little attention to the finer points of Chinese social interaction. He treated the Chinese as he would any employee: demanding excellence, refusing to hear excuses, and chastising failure. Wang and Fortune would travel together on and off for years, and the servant valiantly tried to negotiate the workings of *guanxi* on his master's behalf. Wang effectively created Fortune's identity as a mandarin by forging a fictitious network of prestigious connections for him, elevating his master's face (and, not incidentally, elevating his own status by association). He also bribed and negotiated on Fortune's behalf, not just for favors but for face.

It might have seemed to Fortune that these obligations only served to increase Wang's profits and interfered with the efficiency of the expedition. Although Fortune had traveled in China before, he remained an easy mark for those who were entrusted with keeping his secrets. Out of necessity he relied on his servants, making a family of his travel companions: It was a network of consideration, if not care. He did not appreciate the high level of face status he conferred upon them by depending on them so heavily. Nor did he foresee how rebellious this would make them.

After the trip through Hangzhou, Wang engaged a cargo boat to take Fortune out of the city toward the tea regions of Anhui Province, to the rich Yangtze valley. Fortune's berth would be in the stern, next to a dwarf's; he would receive a straw mat to sleep on (the servants slept abovedecks) and a basin of hot water with which to wash each morning (unlike Europeans, nineteenth-century Chinese bathed daily). The cost of his passage would include two meals of rice gruel and one of rice per day. Under Fortune's sleeping mat were two coffins—presumably occupied. Such was the reverence in China for ancestors and ancestral graves that merchants from the provinces who had the misfortune to die in the coastal cities, away from their clans, were always repatriated to their home villages. It was believed that otherwise their ghosts would be confused when the time came to bestow good fortune on their relatives and that they would become angry and therefore vengeful at finding themselves buried so far from home. Heedless of this custom, Fortune assumed the containers were simply a bed and slept soundly on them.

The river towns around Hangzhou were eerie and dilapidated, their crumbling walls overrun with weeds and shrubs. These small settlements were dens of robbers and pirates waiting to prey on the newfound wealth sailing past toward the China Sea. His boatmen regaled Fortune with stories of local terror and brigandry, which "almost made me believe myself to be in dangerous company," he recalled. Fortune posted the coolie to keep watch over his cabin every evening, augmenting the work of the night watchman. "How long these sentries kept watch I cannot tell, but when I awoke, some time before the morning dawned, the dangers of the place seemed to be completely forgotten, except perhaps in their dreams, for I found them sound asleep. . . . No one seemed to have harmed us during our slumbers."

However much the boatmen might have enjoyed spinning a frightening yarn, maintaining a good relationship with the crew

was an essential lifeline for the three travelers. Boatmen negotiated the river routes at their own discretion and were accustomed
to doing so, for there were always hostile forces sailing on the waterways of China, whether arms and opium dealers or rebels and
religious renegades. Above all other considerations was the fact
that the boatmen had to be kept content lest they betray their illegal human cargo. If the sailors did not know the precise nature
of Fortune's secret, they took it for granted that every man traveling in China had one.

While the river junk was docked in Hangzhou, waiting for a
full roster of passengers before pushing off, Wang conducted
some business dealings of an indeterminate nature with the head
boatman. Perhaps he was scheming for a larger fraction of the
cost of the passage to find its way back to his pocket—with a split
for the boatman, naturally. It is possible he had arranged for a
false "chop"—the receipt of purchase that served as a ticket—so
that Fortune would never know the full price of his passage. Or
maybe Wang had found some cargo downriver that commanded
a higher price upriver and was bartering for a little extra stowage
to be added to Fortune's bill. Whatever his game, Wang owed a
significant amount of money to the captain.

He paid his debt to the boatman in silver, using a silver dollar, but on shore that night, when the captain was drinking,
gambling, and otherwise enjoying himself before resuming the
long journey, he put down Wang's dollar to pay for his evening's
entertainment. The innkeeper examined the coin and told him it
was counterfeit and unacceptable. Mexican silver dollar coins
were rare inland, although they were the accepted currency of
the international settlements on the coast—the very first world
currency, in fact. (Mexico received independence from Spain in
1821, but the coins were still considered the "principal money" of

the China trade throughout the century.) The foreign trade that entered the country after the Opium Wars had increased the number of silver dollars in circulation in China so that even a simple boatman would likely have been aware of silver's high value against local currency—especially since the copper coinage of China was in the process of being radically devalued. Yet the relative scarcity of silver inland would have meant the captain might not have known how to identify a true Mexican dollar.

Settling his debts at the gaming table in Chinese coin, the captain then returned to his ship, where Wang adamantly refused to take the dollar back. A dollar is a dollar, he insisted. The captain was equally resolute: He wanted a refund. It was late at night, Wang was drunk, and the two men continued to argue. How was the captain to be recompensed? How could Wang be certain that this was even the same dollar he had handed over that morning? When the argument grew heated, the captain finally threatened to alert the mandarin that his servant was passing false coin.

At last Wang agreed to pay the captain in Chinese money instead, mumbling a protest—"That dollar was good enough"—as he threw down its equivalent in heavy strings of copper coins, which were the common medium for local exchange.

The boatman was satisfied with the settlement at first, but when he counted his haul, he found it was short. Wang was still trying to take advantage of him.

Wang exploded: "I gave you a dollar, and you said that was bad. I changed it and gave you copper cash, and you return them. Pray, what do you want?"

By now Fortune and other passengers had gathered around in the cool night air, drawn by the increasing volume of the brawl. The boatman finally left the scene resentfully, clutching the cash that he insisted felt far too light. Although Fortune could not follow the angry discussion between the two men, he began to realize the liabilities of traveling with someone like Wang, and his suspicions would only grow.

As he traveled inland, Fortune became distracted by the scenery of China's rich agricultural regions. The fields were terraced and tawny in their harvest colors. Orchards abounded. The peaches and plums had been gathered, but apples still hung heavy on the boughs, while oranges were just now ripening. Fortune was beguiled, a farmer's son returning to the rural life he remembered, but at the same time he was galvanized by the prospect of being a frontiersman, setting foot on virgin ground. From the safe distance of the boat, he admired the industry of the Chinese farmers. Tucked into every patchwork display of cereal crops—wheat, rice, and corn—was one evergreen patch of tea. Every farm or household had a postage-stamp-size tea garden of its own, each a sign to Fortune that he was nearing the intended target.

Apart from the palpable unease that remained between Wang and the captain, the river journey was a happy one for Fortune. As they advanced farther into the interior, he encountered the wildest terrain he had yet seen. Nothing contents a naturalist more than watching an unblemished scene unfurl, like a scroll, as his boat navigates the twisting course of a river. Fortune would see the occasional pagoda or temple nestled in the hills, announcing a nearby town, but mainly he saw craggy mountains, waterfalls, and the lush bamboo-covered vistas of the Far East.

When the rapids were fast or the riverbed shallow, the boat proceeded alongside the riverbank, where a barefoot team of fifteen near-naked coolie trackers slung ropes over their shoulders and hauled the vessel along. At other points on the river the sailors poled their flat-bottomed boat around dangerous rocks midstream. In such instances it took hours to make any progress. While other travelers would take a nap or pull out the tiles and start the evening's gambling until mealtime interrupted the languor of the day, Fortune took advantage of the slow passage to roam the adjacent hillsides and collect plant samples. Early each morning he and his two men would climb a nearby hill to take a sighting of the curve of the river ahead. If it was a day on which the boat would make little headway, Fortune would stay onshore and catch up with the vessel later in the afternoon. One can only wonder what Wang and the coolie made of their master's kit: picks and shovels, blotting paper, notebooks, magnifying glass, specimen jars, Ward's cases, and wicker baskets. It was their first exposure to his obsessive note taking, sample digging, and aimless wandering, all apparently with the purpose of collecting a puzzling variety of plants and shrubs, which surely would prove to be more trouble than they were worth. Fortune lived for these interludes. "The weather was delightful, the natives quiet and inoffensive, and the scenery picturesque in the highest degree. . . . My Chinamen and myself, often footsore and weary, used to sit down on the hilltop and survey and enjoy the beautiful scenery around us. The noble river, clear and shining, was seen winding amongst the hills; here it was smooth as glass, deep, and still, and there shallow, and running rapidly over its rocky bed." Were the days so pleasant for Wang, when he spent and made no money? Or for the coolie, struggling behind with armloads of hardware and heavy glazed cases hanging off his shoulders? The hillside jaunts inspired Fortune in no small part because they might provide the means of his advancement. In his first few days out of Hang-zhou, he discovered the hemp palm, *Trachycarpus fortunei*, which he would send off to the Royal Botanic Garden at Kew upon his return to Shanghai and which would survive its first English winter in 1850. On another of these early forays he also found a remote private garden where he came upon the very picture of mournfulness, the Chinese weeping or funeral cypress, *Cypressus funebris*.

"What a fine tree this of yours is!" he told the gardener who stood nearby. "We have never seen it in the countries near the sea where we come from. Pray give us some of its seeds." (While the cypress would ultimately be raised in Britain, it was a tender tree that did not take well to Kew and was therefore never a sought-after item at auctions.)

China had already proven its value in enhancing Britain's gardens—value that Fortune himself had played no small part in establishing. Housing the richest collection of temperate flora in the world, China is the source of many of the plants that are now familiar elements in our landscapes. Its ornamental floral descendants grace our springtimes with bold yellow forsythias and splashy rhododendrons, azaleas, and camellias; our summers with roses, peonies, gardenias, clematis, apricots, and peaches; our autumns with chrysanthemums; and our winters with citrus—oranges, grapefruit, and lemons.

For his herbarium Fortune also collected a catalog of pressed plants that he dried and maintained for the benefit of his fellow botanists. He trained his servants carefully in this work: constantly changing sheets of blotting paper and keeping a careful eye out for mold, insects, and other things that might foul a collection. "It is possible for an intelligent native to do a certain amount of the changing of the drying paper," wrote an expert on plant hunting in China, "but the arranging of the plants in the press on the first occasion may make or mar the beauty of the specimens." Fortune also collected specimens for the Wardian cases, carefully shaking dirt from roots, placing samples in boxed soil, watering them, and then sealing the glass, in the hope that the tender uprooted plant would take to its new environment and survive long enough to make it to Shanghai and then to England.

Beyond the active work of digging, pressing, replanting, and shipping, Fortune took care to honor his obligation to history by keeping a scrupulously meticulous notebook. The vast treasure trove of his memoirs from China contains reams of minute botanical detail that testify to his skill as a collector, not to mention as a businessman. Of the fifteen thousand plant species in China, nearly half of which were endemic, Fortune sampled and cataloged so many ornamentals that there was almost nothing left for future plant hunters to exploit. Those who did follow Fortune in China would have to head west, toward Yunnan and up into the Himalayas, to find their own terra incognita.

As the boat continued upstream, the other passengers' behavior toward Fortune began to change subtly: His fellow travelers no longer addressed him by his Chinese name, Sing Wa. In fact, Fortune noticed that they were no longer addressing him at all. Instead, his shipmates stole glances at him, muttering just outside the range of his hearing, and generally sought to avoid him. He seemed to have become an object of general intrigue, even more so than the dwarf.

Their attention made him uncomfortable just when he had begun enjoying his disguise. Fortune had come to believe that he

was losing his foreignness, that his British persona was something he could shed like a change of clothing. He had become
convinced that there was nothing innate about his outsider status. Although he was not yet completely comfortable with using
chopsticks under the watchful eyes of Chinese travelers, he had
been growing more so. At mealtimes he sensibly kept his distance from other diners, who sat drawing on long tobacco pipes
while drinking distilled grain alcohol, growing louder and more
belligerent with every toast. He noticed that he was beginning to
understand others as they mumbled drunkenly, the foreign sylla-
bles shaping themselves into words and the words into meaning-
ful sentences. He did not join in general conversation, but his ear
was becoming attuned to the local language. Now, however, it
suddenly seemed to him that the other passengers had seen
through his ruse.

Perplexed, he finally called Wang aside to ask what had
brought about this change in his shipmates' attitude.

Wang explained that the coolie had once again blown Fortune's cover, that in a flurry of resentment and angling for "face"
he had unmasked his master.

"That coolie, he too much a fool-o; he have talkie that you
no belong to this country; you more better sendie he go away,
suppose you no wantye too much bobbly." He's making trouble, Wang insisted in pidgin. He has put you in danger. Get rid
of him.

Fortune, who remained ignorant of the finer shadings of face,
did not appreciate how the betrayal might have benefited the
coolie within the social context of the boat, where he craved the
status he could not get from his master. In the end, however, giving up Fortune's secret did the coolie little good, for the crew
now seemed to consider that the broken confidence was a punish-

able offense that left him in their power. Fortune himself was furious with his servant, and the hapless coolie grew more surly and miserable with each day.

When the ship moored at night, its wooden hull creaking as it bobbed gently in the water, a guard was typically posted to patrol. "The boatmen informed me that this part of the country abounded in thieves and robbers, and that they must not all go to bed at night, otherwise something would be stolen from the boat before morning," Fortune recalled. But on the humid and moonless night following Wang's contretemps with the captain, no one kept watch.

"Wake up!" Wang whispered, shaking Fortune awake. *Qichuang!* The boatmen had all gone into town while the passengers slept onboard. Wang believed there was a conspiracy afoot against Fortune. The night watchman had been dismissed, and the boatmen were planning to kill them all. He had heard rumors of such a plot from other passengers or whispers between the crew members.

"They have now gone into town to get some of their friends to assist them," Wang insisted. "They are only waiting until they think we are fast asleep."

Fortune rose from his mat to look through the porthole toward the shore and saw a distant string of lanterns that seemed to be moving toward them. If he had indeed been identified as a foreigner, hundreds of miles from a sanctioned treaty port in violation of the strictures of the Treaty of Nanking, the local authorities would never intervene on his behalf.

"Get up! Get up! Quick, quick!" Wang said. Chop-chop!

Fortune remained in his cabin with "all the composure I could command" as the lanterns drew nearer and what was now clearly the crew approached the boat. He took his place by the cabin door, anticipating the worst. In his account he makes no mention of where his guns were—or whether he was even armed that night. "My two Chinamen appeared in a state of great alarm and kept as close to me as they possibly could," he recalled.

The first member of the band entered the cabin and found Fortune and his servants awaiting him. The boatman looked sheepish, as if he had not expected to find his quarry awake.

The intruder broke into an ingratiating grin, then shrugged and said nothing at all. He turned his back on Fortune and left. He wanted nothing.

Other passengers were awake by now, but Fortune could not make sense of what was happening.

"Now you see that?" Wang insisted. "You would not believe me when I told you that they intended to seize and drown us, but had we not been awake and fully prepared, it would have been all over with us."

Fortune, in fact, had no idea how great the threat had been or even whether he was in any danger at all. What had happened, after all? A man had opened a door and then walked away. Would events have transpired differently if he had been asleep?

The captain of the vessel returned later that night, acting as if nothing had happened.

Perhaps nothing had. After the general commotion onboard, Fortune found he could not rest. "Cold and sleepy," he listened to the nearby clank and squeak of waterwheels that powered the primitive mills by the river. For a man who prided himself on his powers of observation, it was unnerving not to understand what had transpired or know whether or not he had faced a legitimate threat. Fortune was coming to understand just how much he depended on his companions: He could not trust them, but he had no alternative.

6

A Green Tea Factory, Yangtze River, October 1848

With Wang walking five paces ahead to announce his arrival, Fortune, dressed in his mandarin garb, entered the gates of a green tea factory.

Wang began to supplicate frantically. Would the master of the factory allow an inspection from a visitor, an honored and wise official who had traveled from a far province to see how such glorious tea was made?

The factory superintendent nodded politely and led them into a large building with peeling gray stucco walls. Beyond it lay courtyards, open work spaces, and storerooms. It was warm and dry, full of workers manufacturing the last of the season's crop, and the woody smell of green tea hung in the air. This factory was a place of established ceremony, where tea was prepared for export through the large tea distributors in Canton and the burgeoning tea trade in Shanghai.

Although the concept of tea is simple—dry leaf infused in hot water—the manufacture of it is not intuitive at all. Tea is a highly processed product. At the time of Fortune's visit the recipe for tea had remained unchanged for two thousand years, and Europe had been addicted to it for at least two hundred of them. But few in Britain's dominions had any firsthand or even secondhand information about the production of tea before it went into the pot. Fortune's horticultural contemporaries in London and the directors of the East India Company all believed that tea would yield its secrets if it were held up to the clear light and scrutiny of Western science. Among Fortune's tasks in China, and certainly as critical as providing Indian tea gardens with quality nursery stock, was to learn the procedure for manufacturing tea. From the picking to the brewing there was a great deal of factory work involved: drying, firing, rolling, and, for black tea, fermenting. Fortune had explicit instructions from the East India Company to discover everything he could: "Besides the collection of tea plants and seeds from the best localities for transmission to India, it will be your duty to avail yourself of every opportunity of acquiring information as to the cultivation of the tea plant and the manufacture of tea as practised by the Chinese and on all other points with which it may be desirable that those entrusted with the superintendence of the tea nurseries in India should be made acquainted." But the recipe for the tea was a closely guarded state secret.

In the entry to the tea factory, hanging on the wall, were inspiring calligraphic words of praise, a selection from Lu Yu's great work on tea, the classic *Cha Ching*.

The best quality tea must have
The creases like the leather boots of Tartar horsemen,
Curl like the dewlap of a mighty bullock,
Unfold like a mist rising out of a ravine,
Gleam like a lake touched by a zephyr,
And be wet and soft like
Earth newly swept by rain.

Proceeding into the otherwise empty courtyard, Fortune found fresh tea set to dry on large woven rattan plates, each the size of a kitchen table. The sun beat down on the containers, "cooking" the tea. No one walked past; no one touched or moved the delicate tea leaves as they dried. Fortune learned that for green tea the leaves were left exposed to the sun for one to two hours.

The sun-baked leaves were then taken to a furnace room and tossed into an enormous pan—what amounted to a very large iron wok. Men stood working before a row of coal furnaces, tossing the contents of their pans in an open hearth. The crisp leaves were vigorously stirred, kept constantly in motion, and became moist as the fierce heat drew their sap toward the surface. Stir-frying the leaves in this way breaks down their cell walls, just as vegetables soften over high heat.

The cooked leaves were then emptied onto a table where four or five workers moved piles of them back and forth over bamboo rollers. They were rolled continuously to bring their essential oils to the surface and then wrung out, their green juice pooling on the tables. "I cannot give a better idea of this operation than comparing it to a baker working and rolling his dough," Fortune recalled.

Tightly curled by this stage, the tea leaves were not even a quarter the size they had been when picked. A tea picker plucks perhaps a pound a day, and the leaves are constantly reduced through processing so that the fruits of a day's labor, which filled a basket carried on a tea picker's back, becomes a mere handful of leaves—the makings of a few ounces or a few cups of brewed tea. After rolling, the tea was sent back to the drying pans for a second round of firing, losing even more volume at every contact with the hot sides of the iron wok.

With leaves plucked, dried, cooked, rolled, and cooked again, all that was left to do was sort through the processed tea. Workers sat at a long table separating the choicest, most tightly wound leaves—which would be used in the teas of the highest quality, the flowery pekoes—from the lesser-quality congou and from the dust, the lowest quality of all.

The quality of tea is partly determined by how much of the stem and rougher lower leaves are included in the blend. The highest-quality teas, which in China might have names like Dragon Well, or in India FTGFOP1 (Finest Tippy Golden Flowery Orange Pekoe First Grade), are made from the topmost two leaves and the bud at the end of each tea branch. The top shoots taste delicate and mild, and are only slightly astringent; therefore the most pleasant and refreshing.

The distinctive quality of tea comes from essential oils that leach flavor and caffeine into a cup of hot water. These chemical compounds are not necessary for the primary survival of the tea plant's cells; they are what is known as secondary compounds. Secondary chemicals help plants in many different respects, such as defending them against pests, infections, and fungus, and aiding them in their fight for survival and reproduction. Tea, like other green plants, has several defense systems against predators: Caffeine, for instance, is a natural insecticide. Almost all of tea's thick waxy leaves, apart from the topmost shoots, are bitter and leathery and difficult to bite through. Tea also has hard, fibrous stalks to discourage animal incursion. Clumsy pickers can compromise the quality of tea by including a leaf farther down the stem and even some of the stem itself; this will make for a harsher, more tannic brew, and in China it will be qualified by names suggesting crudeness, such as dust.

The workers sat at long low tables to pick through the leaves and sort out any pieces of stem. They also looked for any insects that might have tainted the batch, as well as small stones and pieces of grit from the factory floor. Even with a measure of quality control, tea was not a clean product in any sense, which is one of the reasons that Chinese tea drinkers traditionally discard the first cup from any pot. "The first cup is for your enemies," the saying goes among connoisseurs.

Culinary historians know nothing about who first put leaf to water. But where human knowledge has failed, human imagination has inserted itself. Many Chinese believe that tea was discovered by the mythical emperor Shennong, inventor of Chinese medicine and of farming. The story goes that one day the emperor was reclining in the leafy shade of a camellia bush when a shiny leaf dropped into his cup of boiled water. Ripples of light green liquor soon began to emerge from the thin, feathery leaf. Shennong was familiar with the healing properties of plants and could identify as many as seventy poisonous plants in a daylong hike. Convinced that the camellia tisane was not dangerous, he took a sip of it and found that it tasted refreshing: aromatic, slightly bitter, stimulating, and restorative.

Ascribing the discovery of tea to a revered former leader is a characteristically Confucian gesture—it puts power in the hands of the ancestors and links the present day to the mythic past. But Buddhists in China have their own creation story for tea, featuring Siddhartha Gautama (Gautama Buddha). As a traveling ascetic, legend tells us, the young monk Siddhartha was wandering on a mountain, perfecting his practice, and praying without ceasing. The weary supplicant sat down by a tree to meditate, to contemplate the One and the many faces of redemption, and promptly fell asleep. When he awoke, he was furious at his own physical weakness; his body had betrayed him, his eyes were leaden, and drowsiness had interfered with his quest for Nirvana.

In a fit of rage and determined that nothing would again impede his path to Truth and Enlightenment, he ripped out his eyelashes and cast them to the wind, and in all the places they fell sprang forth a fragrant and flowering bush: the tea plant. Indeed, the fine, silvery down on the undersides of the highest-quality tea leaves resembles delicate eyelashes. Buddha, all great and compassionate, bequeathed to his followers a draft that would keep them aware and awake, invigorated and focused, an intoxicant in the service of devotion.

Before Fortune, botanists had failed in their attempts to decode the formula for tea. His first collecting trip to China in 1843, for the Royal Horticultural Society, had taken him to the fringes of tea territory as part of his general collecting mandate. At that time he had made an important discovery: Green tea and black tea came from the same plant.

The Linnaean Society had hitherto declared unequivocally that green and black tea were siblings or cousins, closely related but under no circumstances twins. The great Linnaeus, a century before, working from dried samples brought back from China by earlier explorers, concluded that the two were distinct taxa: *Thea viridis* and *Thea bohea*. *Thea viridis*, or green tea, was said to have alternating brown branches and alternating leaves: bright green ovals that were short-stalked, convex, serrated, shiny on both sides, and downy beneath, and with a corolla, or flower, of five to nine unequally sized white petals. *Thea bohea*, black tea, was described as looking nearly the same—only smaller and somewhat darker.

On his first trip Fortune expected to find identifiable black tea plants in gardens known to produce black tea. Yet he discovered that the tea plants there looked just like the green tea plants in the green tea gardens. Over the course of that first three-year visit, when procuring several tea samples and thoroughly investigating them, he had concluded that any difference between green tea and black was the result of processing alone. His botanical colleagues were slow to agree, requiring more proof.

Black tea is fermented; green tea is not. To make black tea, the leaves are allowed to sit in the sun for an entire day to oxidize and wilt—essentially to spoil a little. After the first twelve hours of stewing, black tea is turned, the liquor is stirred around, and the mixture is left to cure for another twelve hours. This longer curing process develops black tea's tannins, its strong bitter flavor, and its dark color. Although it is called fermenting, the process of making black tea is technically misnamed. Nothing ferments in a chemical sense; there are no microorganisms breaking down sugars into alcohol and gas. Black tea is, rather, cured or ripened. But the language of wine colors the language of all beverages, and so the label of "fermentation" has stuck to black tea. (Indeed, if tea does ferment and fungus grows, a carcinogenic substance is produced.)

Given that to that point no European botanist had seen tea growing or evaluated it in its living state, the Linnaean Society's confusion on the subject is understandable. Fortune's documentary evidence ultimately changed tea's Linnaean classification. It would soon be known categorically as *Thea sinensis*, literally tea from China. (Later still it would be reclassified as part of the Camellia family, *Camellia sinensis*.)

As he made his way through the green tea factory, Fortune took note of something both peculiar and more than a little alarming

on the hands of the tea manufacturers. It was the kind of observation that, once reported, would be an invaluable boon to the
burgeoning Indian tea experiment, with the power to boost the
sales of Indian tea over Chinese. While staring at the workers
busy in the final stages of processing, he noticed that their fingers
were "quite blue."

Among the blenders and tasters of the London auction it was
generally assumed that the Chinese engaged in all manner of duplicity, inserting twigs and sawdust into their teas to bulk up the
loose leaves. It was said that the Chinese were brewing their own
breakfast tea, saving the soggy leaves to dry in the sun, and then
reselling the recycled product as fresh tea for the gullible "white
devils." There was no trust in the trade, no faith in the goodwill
of the Chinese manufacturers.

But the blue substance on the fingers of the Chinese workmen
seemed to Fortune a matter of legitimate concern. What could be
the source of this? He and others had long suspected that the
Chinese were chemically dyeing tea for the benefit of the foreign
market. He was now in a position to prove or disprove the
charge.

He watched each step of the processing carefully, saying nothing, making notes, and occasionally asking Wang to put a question to a manager or worker. At one end of the factory the
supervisor stood over a white porcelain mortar. In the bowl was a
deep blue powder, made finer and finer with each grind of the
pestle. The superintendent was in fact preparing iron ferrocya-
nide, a substance also known as Prussian blue, a pigment used in
paints. When cyanide is ingested, it binds to iron inside cells, in-
terfering with the absorption of certain enzymes and compro-
mising a cell's ability to produce energy. Cyanide affects the
tissues most needed for aerobic respiration, the heart and lungs.

In high doses cyanide can bring on seizures, coma, and then cardiac arrest, killing quickly. At lower doses cyanide leads to weakness, giddiness, confusion, and light-headedness. Exposure to even low levels of cyanide over long periods of time can lead to permanent paralysis. Fortunately for the tea drinkers of Britain, Prussian blue is a complex molecule, so it is almost impossible to release the cyanide ion from it and the poison passes harmlessly through the body.

Elsewhere in the factory, however, over the charcoal fires where the tea was roasted, Fortune discovered a man cooking a bright yellow powder into a paste. The smell was terrible, like that of rotten eggs. The yellow substance was gypsum, or calcium sulfate dehydrate, a common component of plaster. Gypsum produces hydrogen sulfide gas as it breaks down. While the gas is produced naturally by the body in low doses, in high doses it acts as a broad-spectrum poison, affecting many of the body's systems simultaneously, particularly the nervous system. At lower concentrations gypsum acts as an irritant; it reddens the eyes, inflames the throat, and causes nausea, shortness of breath, and fluid in the lungs. Consumed over the long term it might produce fatigue, memory loss, headaches, irritability, and dizziness. It can even induce miscarriage in women, and failure to thrive in infants and children.

Fortune estimated that more than half a pound of plaster and Prussian blue was included in every hundred pounds of tea being prepared. The average Londoner was believed to consume as much as one pound of tea per year, which meant that Chinese tea was effectively poisoning British consumers. The additives were not included maliciously, however, for the Chinese simply believed that foreigners wanted their green tea to *look* green.

"No wonder the Chinese consider the natives of the West to be a race of barbarians," Fortune remarked.

But why, he asked, were they making green tea so extremely green, since it looked so much better without the addition of poison and since the Chinese themselves would never dream of drinking it colored?

"Foreigners seemed to prefer having a mixture of Prussian blue and gypsum with their tea, to make it look uniform and pretty, and as these ingredients were cheap enough, the Chinese [have] no objection to [supplying] them as such teas always fetch . . . a higher price!"

Fortune surreptitiously collected some of the poisonous dyes from the factory, bundling them up in his wax-dipped cloth sacks and stowing them away in the generous folds of his mandarin costume. As a scientist he wanted samples to analyze, but most of all he wanted to send additional ones back to England.

These substances would be prominently displayed in London's Great Exhibition of 1851. In the glittering Crystal Palace, Britain displayed to the world all its industrial, scientific, and economic might, including the green tea dyes. This public exhibition marked the moment when tea, the national drink of Britain, came out of the shadows of myth and mystery and into the light of Western science and understanding. Fortune unmasked unwitting Chinese criminality and provided an irrefutable argument for British-manufactured tea.

7

House of Wang, Anhui Province, November 1848

Although the day's light was fading, Fortune could see twisted pines penetrating a sea of cloud, ornamenting the sharp outlines of the hills beneath. The landscape might have been made for the gestural strokes of a scroll painter's brush. It was no wonder, Fortune thought, that so much of China's appreciation for tea was echoed in its arts, painting, pottery, and poetry, for who would not wish to reproduce, savor, and preserve forever such intense beauty?

The Wang residence was a mere two miles from the foot of the steep slopes of Sung Lo Mountain, and its proximity to such famous tea grounds may explain why Fortune had kept Wang as his principal guide, despite his grasping ways and proclivity for trouble. He came from people for whom tea-growing was a native art. Returning to the tea peaks from the coast was a ritual that Wang's ancestors had performed for centuries.

Wang strode through the heavy wooden doors of his childhood home steps ahead of Fortune. He entered without carrying any baggage—prince of the palace, happy to be home, eager to announce the arrival of the distinguished foreign mandarin, and, not coincidentally, Wang's own recent stroke of good luck. Given the remoteness and poverty of the area, and the corresponding lack of public inns, Fortune had agreed to make his residence at the Wangs' home.

His parents embraced Wang with joy. His mother doted on him and asked if he had already eaten his rice, while his father beamed with pride. They were also appropriately surprised and impressed by the novelty and stature of the mandarin to whom their son was attached, protesting earnestly that their home was not worthy of such an honor. Looking around, Fortune had to agree, if only silently. The door through which he had just entered featured hardware so crude and primitive that any blacksmith in England would have been ashamed to call it a latch or hinge. Strips of faded red paper hung limply over the lintel, to bring blessings on the family and protect the house from evil.

At first glance Fortune could see little opportunity for comfort within the Wang household. The house itself was rickety and almost physically impossible: a dwelling perched so precariously on the side of a cliff that it stood as a monument to the persistence of the laborers who had built it there. Fortune had seen many drab dwellings on his travels but had not as yet dared to enter a peasant's home. The beaten earth walls were thick and painted white, in stark contrast to the blackened roof timbers. The roofline was lavishly decorated and ornately carved: Upturned corner tiles had the outlines of animals pressed into the clay to frighten dangerous spirits away—it was believed there were demons everywhere. The house had small windows covered with rough lattice-wood screens to keep out the birds but not the vermin (and most certainly not the flies).

The Wang home was too small to provide separate quarters for men and women, an accommodation that Fortune must have come to expect from his visits to the gardens of great men where he had collected some of his finest specimens on his previous trip to China. In the 1,000 or so square feet of the Wang household, the respective territory of the sexes was as clearly defined as if they had been physically separated. The men occupied the public space, a sizable room with sacks of rice piled in one corner. The occasional chicken wandered through, and a few select pieces of furniture—a wedding chest, a cane kitchen cabinet, and a bent-wood chair—bore testament to better days. Off this were other, smaller rooms where the women lived, babies were nursed, food was prepared, and cloth was woven.

Yet Fortune was determined not to be outdone in showing the usual courtesies, and almost immediately he seemed to be among the best of friends. Wang Senior was a farmer and, like many Chinese, at the mercy of the country's growing pains, the boom-and-bust cycle of an agrarian economy undergoing a population explosion. He had known prosperity, only to be brought low by famine and hard times. Heedless of his poverty, the elder Wang was generous to Fortune, the foreigner. Soon a great dinner was set out with the best that the family could offer, and the fattiest cut of pork, the first taste of the stew, was offered to their distinguished guest. After dinner they retired to their cramped chambers, with Fortune intending to make an early start on the trek up Sung Lo Mountain to commence collecting seeds.

In the morning, however, rain was falling in torrents, and it was clear that they would have to stay indoors. It was then that Fortune became acutely aware that the home he was staying in held not one but four separate families—various branches of the larger extended clan, each with its own children, kitchen, and stove. The hours leading up to mealtimes proved especially torturous for Fortune. With four kitchen fires blazing and no chimneys whatsoever, the foul smoke and smell of burning pork fat filled the structure, wafting into every nook and cranny, dirtying anything that was clean. Fortune's eyes stung and watered terribly, but to his amazement the other dwellers in the Wang home took this and the myriad other discomforts of cohabiting and collectivism in their stride. He supposed they simply knew no other way of life.

Despite being a farmer, and an impoverished one at that, the elder Wang was a literate man. China has always enjoyed a high level of peasant literacy—it had a movable type printing press almost four thousand years before Gutenberg printed his first Bible. The elder Wang had trained to be a scholar and poet, as many young men of the region were despite their meager circumstances. When families in the neighborhood became successful merchants, they endowed a local school or Confucian academy in their village. Their sons were coached to compete in the national civil service exams, so as to join the highest levels of the government and become *jinshi*, or scholars. Sung Lo was well known for producing *jinshi*; in the sixteenth century it was noted that there was one poet for every three merchants in the region. At night, after supper, while the dark and continuing downpour kept the full household indoors, Wang Senior kept alive the ways of his elders by reading Chinese fables to his extended family while they huddled together to keep warm. That single act in itself was a happy sight to Fortune, who up to that point had found little in his servant's pastoral household to please him.

The Wang home bore ample evidence of scholarship, such as the calligraphic poems hanging on scraps of paper on the walls. In China calligraphy is considered an art and is revered by men of standing and education. Because the Chinese language is pictorial, it lends itself to beautified visual renditions. The brush, the ink stick, paper, and ink slabs, known as the Four Treasures of the Scholar's Studio, were all obtained from local forests. Each brush stroke was thought to reflect the character and sentiments of the writer, his psychological state of mind at the time he was contemplating the poetry he inscribed. To be a good calligrapher required training and respect for tradition, and so it complemented the Wang family values of discipline and repect.

Fortune could also see something familiar in the family's existence when he compared with his own modest roots. He noted that "the Chinese cottages, amongst the tea hills, are simple and rude in their construction, and remind one of what we used to see in Scotland in former years, when the cow and pig lived and fed in the same house with the peasant. Scottish cottages, however, even in those days, were always better furnished and more comfortable than those of the Chinese at the present time." And where Scotland had whisky, in Xiuning County there was tea. Like Scotch, tea was produced where other crops would not thrive. And as with Scotland's finest malts, it was in these poor cottages that the best teas with the most curious names were found.

In the hierarchy of Chinese life, tea was ranked as one of the seven necessities, along with firewood, rice, oil, salt, soy sauce, and vinegar. For the Wang family to participate in the manufacture of life's basic needs was an honor; they saw themselves as meeting the needs of the broader world order. Although tea was a necessity, it was also considered a luxury. It took time to enjoy tea and money to buy it—if you weren't growing your own. It was the greatest joy of the official classes to sit and drink tea while writing poetry. The Wangs and millions of families like them made such civilized pleasures possible.

While the steep mountainsides of Sung Lo were ideal tea grounds with their thick mists, well-drained soil, and indirect sun, those same conditions made growing other crops a human chore. "The district is [set] among 10,000 mountains. Its land is difficult and not flat. Its earth is tough and unchanging. . . . Though the people are industrious and use all their strength, the harvest is only enough to provide for half [the population]," wrote a local mandarin in an 1815 county gazetteer. "Where there is land lacking, those who support [themselves] by tea-growing are seventy to eighty per cent. From this they clothe and feed themselves and pay their land and labour taxes," records state. The Wangs worked the land both privately and collectively, tilling a public field to pay for taxes and tribute as well as their private plot to feed their own kin. This division of responsibility was entirely Confucian: The basic family unit was coextensive with a wider social world; the men studied, the women worked, and the mandarins—China's politician-scholars—collected taxes in the name of the emperor.

Sung Lo's slopes had been carved into a series of laddered fields for planting grains such as rice and barley; vegetables such as beans, sesame, squash, eggplant, turnip, onions, bamboo shoots, ginger, and garlic; and fruit and nuts such as peaches, watermelon, papaya, walnuts, and peanuts. The terraced slopes were a marvel of human muscle, a compelling demonstration of what China's giant workforce could accomplish over generations. Even so, many from their region had left farming for trade. "Because agriculture is not sufficient to feed the people of the county, most people are engaged in commerce as their constant business. . . . They traveled to the south and the north. Some were pedlars and some set up their shops. They consider what is abundant and what is in shortage, and buy or sell out according to the trend of demand and supply," reported the local gazetteer three hundred years earlier. The younger Wang, like so many of the young men of the district, had been forced by economic circumstances to leave the tea mountains and try to improve his fortunes in the cities on the coast. "The bitterness you are eating is what makes a man into a man" went one local folk song.

Wang's story was a familiar one throughout China in the nineteenth century. By 1850, the nation had a population of 400 million; for every person alive in 1650 there were three in 1850. As a result of the population boom, villages and towns grew closer together, farms became smaller, and woodlands were mowed down and converted to farmland. With the increasing pressure on land resources, irrigation, fertilizer, and the introduction of New World crops such as corn were used to open up previously marginal farmland. The local ecology suffered; famines, mud slides, and floods became more common. There was competition for basic human resources, yet with manpower so abundant, wages remained low.

The population boom also led to a dearth of marriageable women as female infanticide swept through China. Girls married away from the family house and needed dowries, so they were considered sources of debt. Male children brought honor to a family as well as wives to manage the house and children to carry on the clan name, but the scarcity of marriageable local women and the surfeit of labor offered little incentive for men to remain in their home village. They accordingly migrated to the cities, where they joined gangs or became boatmen or sedan-chair carriers. Wang's area, in particular, was known for producing pawnbrokers. With the new population boom, the established social order was transformed: Men became emigrants while women tried to boost the household economy in any way they could, such as spinning and weaving textiles or, in the Sung Lo region, making and selling their own tea.

The Wangs believed in the Confucian tradition in which there was a hierarchy of occupations: the highest, belonging to scholars and poets, preserved beauty and celebrated order; peasants came second, the cogs in the national machine, growing the food and materials that enabled a complex country like China to function; merchants were ranked at the bottom of the collective heap, earning a living off the hard labor of others but producing nothing of note themselves. While there was great honor in being a peasant, especially a literate one, being unable to support your children was a serious loss of face—and to have their sons working in far-off cities reflected badly on the Sung Lo families.

On the evening of that first rainy day, an odd-looking pair, the giant coolie and the obliging dwarf who had followed Fortune off the boat, arrived bearing the luggage. As they placed the heavy chests and baskets on the stone floor of the entry chamber, a small cloud of dust rose up. The coolie was anxious to describe how he had nearly met his end at the hands of the boatmen Wang had mistreated; he had been forced to take refuge for the night in a temple, the only sanctuary he could find. He reenacted his pursuit in a bizarre pantomime, which Wang did not feel compelled to translate. Fortune watched with amusement, calculating that if the coolie had managed to escape with all the luggage safe and intact, then surely things were not quite as bad as they had been made out to be.

Although his retinue was back at full strength, Fortune had no choice but to wait for the weather to clear. For four days the weather on the mountain remained inclement, which led the Wangs to sleep late and go to bed early, preferring the warmth of their beds to the damp cold of morning. Rain gave hardworking peasants a chance to rest, and it was more than welcome after a long summer and a busy tea harvest. The numerous Wang children were overcome with cabin fever and grew ever more curious about the strange man with his round eyes, long nose, tall bearing, and exotic possessions. Fortune had seldom been confined indoors with his own children for as long as he had with the Wang offspring.

Finally the rain ceased, and, rising to a clear day, Fortune was struck by the beauty of the view from the Wangs' house. According to the Chinese principles of *feng shui*, the luckiest location for a house has a mountain at the back and an open view to the front—easy enough to find in the hilly country of Sung Lo. Like the Wangs' house, most Chinese homes followed the general architectural principles of *feng shui* (which literally translates as "wind and water"), which hold that there are physical laws based on elemental principles to encourage the flow of *qi* (pronounced chi), or energy. Houses and rooms should face south; Chinese dwellings dating back to the twelfth century BC have been found that are aligned along the north-south axis, with a primary entrance facing south. The building must be symmetrical, but with an odd number of bays. Symmetry was revered because it allowed for the presence of a central courtyard, as both a symbol of the center of the home and a place in which to display shrines to the ancestors. A southern outlook was thought to confer an abundance of *yang*, or masculine energy, so the Wang house not only took advantage of the abundance of sunlight and warmth but bestowed upon the family the bounty of auspicious, life-seeking energy. Odd numbers, too, were considered to be *yang*, whereas an even number of rooms would have introduced too much *yin*—or female energy—into the house.

The Sung Lo Mountains were nearly barren except for the tea

plants that clung to their peaks. Although it was famous as the birthplace of green tea and the home of its finest leaves, the region had not been extensively exploited. The production at the time of Fortune's visit was mainly for the use of the growers themselves and the priests whose temples dotted the rugged hillsides.

Fortune quickly prepared to head for the Sung Lo slopes to undertake in earnest the collecting of tea seeds. Although they rose in the distance ahead of him, he might well have been guided by his nose as well as his eyes. It was said of the area that "even without seeing [the mountain], you can smell the tea scent a mile away." Tea is an enormously fragrant plant; its mineral herbal scent permeates the air. It is not surprising that people felt compelled to liberate that aroma, to release its potent sap and spirit. Tea has been called the "essence of mountains" in Chinese poetry, a particularly apposite description in that every ounce of the Sung Lo peak, from loam to limestone, is concentrated within the tea bushes.

As always, Fortune found the hillsides welcoming. His chest expanded in the brisk November air. He bent low to pick off the green seed-filled fruit pods, which looked like little pincushions covered in leathery skin. The bushes were dormant, no longer growing at the fast clip of summer and pushing out new tea shoots. But this was a good time to collect seeds.

The elder Wang followed Fortune every morning, interrupting his meditative walks. However much Fortune wished to pick his seeds and dig his seedlings with a minimum of interference, he could not evade the old man. Fortune did not seek such friendship, but the elder Wang was his self-appointed shadow.

For a week Fortune's days settled into a routine of hard work punctuated by beautiful vistas, new discoveries, and the seemingly continuous negotiations required by the very informal economy of the Wang household. Fortune would rise early, head with Wang and the coolie to the Sung Lo ridge, and gather all the tea seeds they could find. In these magnificent surroundings the work was less exhausting than exhilarating, at least for Fortune. In the evening they returned to the Wang residence and resumed the hard bargaining that seemed to preoccupy and perhaps entertain the residents of Sung Lo.

The coolie, by now convinced that Wang had badly abused him in the course of the disputes with the boatmen, had marked him as a coward—which was reasonable enough, Fortune thought. The coolie therefore insisted that Wang should compensate him for his troubles in the amount of four dollars. Wang, who was on his home ground, thought it safe to ignore this demand. The coolie then intimated that he would get some of his own countrymen to force Wang to pay the amount due, a threat Wang dismissed. When the coolie returned later, unaccompanied, Fortune took him aside and told him firmly that the matter was at an end and that if the coolie pressed his claim any further, Fortune would withhold his pay. This and a small loan to the coolie, which Fortune could never hope to see repaid, seemed to put the matter to rest.

Fortune himself also found it useful to engage in some bargaining. A week or so before reaching Sung Lo, he encountered a barberry, or *Berberis*, a woody shrub with large, glossy, spiny leaves the likes of which were unknown to him but which seemed especially handsome in its bright autumnal colors. It looked as if it would be a good border plant and generally well suited to the European garden. Unfortunately, the one specimen Fortune saw was too large to transplant, and apparently there were no progeny nearby. So, having plucked a leaf of it and marked its location, Fortune initially charged Wang with locating a similar shrub in his home district. When Wang took no interest in this task, it occurred to Fortune that some of the extended Wang family might be enlisted instead.

He showed his sample leaf to a small assembly of Wang's relatives and promised payment to anyone who could bring him a similar plant of transportable size. Much to his delight, within a few minutes one of the assembly returned bearing an entire branch. Fortune compared it to the specimen leaf, confirmed it was a barberry, and asked that the whole shrub be brought to him. This provoked a spirited conversation among the family members. The plant was of medicinal use, they explained, and the owner would not willingly part with it—not for any price. "Sell me this one, and you will be able to buy a dozen others with the money," Fortune implored.

But the man who had found it was unyielding: "My uncle, in whose garden it is growing, does not want money—he is rich enough—but he requires a little of the plant now and then when he is unwell, and therefore he will not part with it." Fortune, sensing an attempt to hold out for a better price, changed tactics. He asked only that they show him the plant, and he promised not even to touch it. He would then bargain directly with the owner. But this also seemed to be too much to expect. Hands flew up and voices squawked in protest until Wang himself intervened, vouching for Fortune to his assembled cousins.

The crew eventually led him to the barberry, giving Fortune the opportunity to negotiate one-on-one with the uncle. It was no use, however. The man insisted that the plant was rare and its berries were of tremendous medicinal value for curing diarrhea, fever, weak appetite, upset stomach, yeast, urinary tract infections, and a whole host of internal ills. He not only flatly refused to surrender the plant but would not even give Fortune a fresh branch for cloning. Fortune did not know whether this was strategy on the uncle's part or whether the old man really did value the barberry as much as he claimed. If all the uncle said about its medicinal abilities was true, it only made the plant hunter want a sample all the more. Was this just a sophisticated ruse? The question was rendered moot when another of Wang's relatives came to him furtively the following day and indicated that he could find Fortune other shrubs of the same sort for the same price. Fortune took him up on the offer, and the young man promptly returned with three fine, healthy specimens. The barberry, it seemed, was common in Sung Lo. Fortune shipped the plants to England, where they became a great favorite for use in hedges and landscape gardens.

In the intimate space of the Wang house, money continued to change hands—Fortune's money, although he had neither consented to nor been informed of the transactions and frequently knew very little about what was happening. The coolie, still peeved by Wang's mishandling of the boatmen, kept trying to extort money from the young translator. Old Wang, believing that no act of hospitality should go unrewarded, argued for hard cash from Fortune's funds in exchange for his room and board. Young Wang, who was by now busy arranging Fortune's passage back to the coast, finalized arrangements that overcharged him by 2000 percent. The squeeze went on and on and on. "Such is the character of the Chinese," Fortune observed drily.

Yet Fortune had nonetheless managed to get the best of the bargain: the most valuable asset in Sung Lo. His servants and hosts might have taken a little for their trouble, but once Fortune left China, he would be taking the finest green tea plants in the world with him, crossing the seas to become prized possessions of the world's only superpower.

Fortune's arrival marked the earliest point of direct contact between the celebrated green tea region and the Western world. In only a few short years his connoisseur's reports of the superior quality of Sung Lo tea would lead to its finding its way into export markets in Europe and America. It was branded Green Tun, and it was to become a great favorite in fashionable society. Although Sung Lo was beset by economic hardship, famine, and poverty, and a rebellion by religious zealots was soon to break out, Fortune's appearance was a turning point for the region's people. The young translator who had brought Fortune to the Sung Lo Mountains was leading his province into the future.

8

Shanghai at the Lunar New Year, January 1849

After visiting three other celebrated green tea districts for seed collecting and after a relatively uneventful trip, Fortune arrived in Shanghai in the days just leading up to Chinese New Year (as determined by the lunar calendar) of 1849. It was the Year of the Rooster, a flamboyant period, according to the Chinese zodiac. The biggest holiday of the Chinese year, it is a time of celebration, of fireworks, of cash gifts and ancestor worship. Expatriates would look forward to it for the fireworks and festivities as well as for the chance to pick up some bargains as the locals scrambled to raise cash that was needed to pay off lingering debts, likewise a New Year's tradition.

Shanghai's lively celebration of the New Year pulsed through the old city's thick walls and eddied out into the foreign concessions. The few Britons in the area, tea and silk traders and Foreign Office men, were easily tempted into watching the dragon dances and the grave-sweeping in the local cemeteries with bewildered interest. The ancient streets were crowded with hawkers, jugglers, and circus performers. Small beggar children with seeping sores tugged at wrists and ankles and offered holiday greetings to all comers—it was good luck to give alms at New Year.

Fortune's first order of business was to post word of his success to the company and the waiting gardeners in India. "I have much pleasure in informing you that I have procured a large supply of seeds and young plants which I trust will get safely to India. These were procured in different parts of the country, some from a celebrated tea farm. . . ." He quickly set to work preparing his booty for shipment to India.

Fortune took up residence in the foreign quarter, again at the home of the trading company Dent, Beale & Co., one of the three leading houses in the Far East. Its revered senior partner, Thomas Beale, "merchant prince and opium mogul," had spent fifty years in the China trade, never returning to Britain. He had already made and lost several fortunes in Cathay. "He was himself one of the old school in its fullest signification: stately in person, somewhat formal, with distinguished manners," wrote a contemporary. Among Fortune's most crucial connections in China was his affiliation with Thomas Dent and his firm. Dent, Beale & Co. owned a compound in the British area, north of noisy Old Shanghai and south of the fetid Souzhou creek, on what had been a towpath for trackers moving boats on the Huangpu River. The compound featured a new and largely empty factory and enough land to provide a garden big enough to accommodate both Fortune's treasures and the amateur projects of the British expatriate gardeners in Dent's circle.

Dent & Co.'s legacy to the international community was an ongoing interest in China's horticulture. For his own amusement Beale kept a garden in Macau, growing the "choicest and rarest" carnations, chrysanthemums, poppies, and all manner of Chinese ornamentals, in addition to a small flock of peacocks and monkeys. Dent & Co.'s local Shanghai gardeners were plant hunting in places where no white man could ever go—and the firm was happy to share its rare cuttings with other European plant aficionados residing in the East.

It is unclear who, outside of the consuls in Hong Kong and Shanghai, was aware of the true nature of Fortune's mission, but if Dent did take any notice of Fortune's tea haul, there was no reason to expect that he was involved in anything other than a botanical study of tea for East India House and the tea traders at Mincing Lane. Dent's main concern was, naturally, maintaining its mercantile success in the Far East. If tea manufacture moved to India, however, the China merchants would lose their most profitable commodity. By helping Fortune, Dent & Co. was inadvertently sowing the seeds of the China trade's destruction.

Fortune had full use of the grounds at Dent's factory in Shanghai to replant and care for his tea seeds and seedlings. The plot occupied the land around some warehouses in an area reclaimed from the rich, silty banks of the Huangpu River, a tributary at the mouth of the Yangtze. This garden itself was not a Chinese-style tableau, intended for leisurely walks and contemplation, but a functional and European one. Its neat rows of seedlings and transplants kept a stock of materia medica at hand and provided the British merchants with food that was recognizable.

Fortune also had access to Dent's Chinese gardener, a knowledgeable man who was armed with all the answers and reasons for whatever gardening choices he made. Beale once told a visitor that "the only way to please a Chinese gardener was to let him do as he pleased," especially when it came to methods "they took much pride in." To interfere with a Chinese gardener at work, Beale understood, was to make him lose face. Nonetheless, Fortune kept his portion of the garden—vast tracts of tea seedlings stretching off into the distance—according to British methods.

With Dent & Co.'s generous support, Fortune's collection

survived transplantation from the wild, but its next relocation would be the most dangerous. Indeed, no plant had yet survived a trip the likes of which he planned for his specimens, sailing from Shanghai to Hong Kong and from there to Calcutta and the hill plantations of the Himalayas. They would contend with heat and sea and salt but also river travel, mountain travel, and monsoons.

Fortune's general anxiety was exacerbated by the biting cold of the Chinese winter. Typically, Dent's garden was a drowsy place where the laborers did not work too hard, especially in the days leading up to the New Year festivities. But Fortune insisted that the gardeners get under way with the critical task of preparing the tea plants for transport. He himself appeared in the garden every morning in gloves and hat, industriously replanting his cuttings and clones, and packaging and labeling seeds and saplings. He was shipping some thirteen thousand young plants to the Himalayas as well as ten thousand tea seeds, about five gallons' worth—not much by volume or even by weight but representing weeks of arduous fieldwork in Wang's sodden tea gardens. As a precaution he divided his hoard, meting out separate packages to be distributed among four separate ships so that if one cargo met with calamity, the others would be unaffected.

Fortune knew that the success of his green tea undertaking was still far from assured at this point. No matter how splendid the seedlings appeared in Shanghai, they might not be able to withstand the dual stresses of both winter and overseas travel, despite the protection of the Wardian case. Moreover, winter is a dangerous time for seeds and shrubs; they become dormant and require special coddling. A good gardener with a nose for frost can take a draft of the night air and tell whether or not plants will face any danger or require swaddling in rice sacks and rags until the sun melts the frost.

Fortune ordered a local glazier and ironworker to construct eleven glazed cases in which to pack his many thousand seedlings. The saplings were perhaps a year old, many much younger, fragile plants with weak and underdeveloped roots. But at least Fortune felt he could rely on his own previous experience of shipping plants by this method. His earlier trip to China had relied heavily on such cases, and the results had left no doubt in his mind as to their efficacy.

The tea seeds presented a different problem. The East India Company had notoriously failed to transport them successfully from China to India over the course of the preceding ten years. In one early shipment, before the widespread use of the Wardian case, all the seeds that had been collected in Canton arrived in the Himalayas dead, delaying the tea-planting timetable for an entire season and incurring huge losses. The tea gardens of the Himalayas remained too small to be profitable largely because of the dearth of quality seeds available, a situation that the specimen plants alone would not alter. From seed to shrub takes six years. If Fortune's efforts failed in 1849, the whole enterprise would once again be delayed a year.

The timing of his harvest was not propitious, however. Typically, the Chinese pick tea seeds in autumn and store them over the winter in baskets of sand until they can be planted in the spring. According to this schedule, Fortune had been late by a month or two in harvesting his seeds, although it seemed reasonable to assume that they would still be viable. However, by the time they arrived in India, the spring planting season would be over. They would arrive in the Himalayas during the monsoon weather when torrents of rain would wash away a gardener's best efforts. These seeds could not be planted until the fall; they would have to stay fresh for over a year, and tea seeds do not keep well.

It was also unclear to Fortune how best to transport the seeds. Standard shipping procedures called for them to be wrapped in paper or cloth sacks. He had earlier been given advice by Dr. William Jameson, the young superintendent of the experimental Himalayan plantations; he had suggested that Fortune try both methods. Fortune was especially thorough in following this advice: He shipped the seeds from four different regions in two different ways. One was a coarse bag containing four paper packages of seeds, and the other was a box of earth mixed with seeds from each region. A third portion of each of the varieties of seeds was kept behind to be sown and grown in Shanghai and then sent out to India once they had germinated into hardier seedlings. Fortune knew that tea seeds were very fragile; they spoiled easily, becoming either waterlogged or dried out in response to the slightest atmospheric change. He might have noticed that many of the seeds from his early collection had not germinated when replanted in the makeshift hothouses of Dent's garden. The best strategy, it seemed, would be to send so great a quantity of seeds that even if most failed, there would still be more than enough to populate the new tea plantations of India.

Fortune sought out Chinese gardeners for their wisdom on the best procedure to store and transport tea seeds. Seeking advice from a native was a daring course for a European in China, not least because the Chinese were reputed to boil or poison tea seeds so that "the floral beauties of China would not find their way into other countries." But Fortune, ever the scientist, boldly approached an old seed dealer, a celebrated merchant named Aching.

"What is the substance you put in the seeds?" Fortune asked

regarding the white ashy matter surrounding them, a mixture that many Europeans thought might be crematorial remains.

"Burnt lice," the old gardener replied.

"Burnt what?" Fortune said, and laughed.

Aching, in his faulty pidgin, repeated, "Burnt lice," this time with "all the gravity of a judge."

"S'pose I no mixie this seed. Worms makie chow-chow he." The ashes were to prevent maggots; the moist climate of China made packed seeds particularly vulnerable to infestation as well as rot. Fortune determined that the old gardener was telling the truth, and there can be little doubt that he experimented with packing green tea seeds in the ashes of burnt rice.

Day after day and into the night, until dusk made it too difficult to dig and the falling temperature turned his fingers numb, Fortune worked on his green tea collections. Thoughts of failure dogged him as he toiled in the mud, but he was a thorough man, and routine calmed him. He made a catalog of each plant and seed: where it was collected and in which case it was sent. He requested equal assiduousness in others. "It will be of great importance if [seeds and saplings] are carefully received and forwarded to their destination," Fortune wrote to the gardeners on the subcontinent. It will "also be very desirable to have a report made upon the condition of the plants and seeds when they arrive in India, which report could be sent to me for my guidance with regard to the number which it will be necessary to collect."

Fortune planned to escort the tea as far as Hong Kong to be certain of its care while in China. "We have no vessel from this place [Shanghai] to Calcutta direct and as any delay or inattention at Hong Kong might prove fatal to the plants I think it much better not to risk committing the [tea] to the care of any person not fully acquainted with such matters." The ignorance of stevedores could destroy Fortune's cargo as easily as weather and the vicissitudes of transplantation.

It would take an entire season for the tea shipment to reach India, and another few months before a letter would arrive from the botanists there informing Fortune whether or not the shipments had been successful. He had no way of knowing where he might be when that important letter arrived—perhaps in the middle of his next trip, to the black tea districts. What would happen if his green tea failed entirely? If the first delivery died or if there were adjustments to be made, he would not even know about it until he returned to Shanghai. He might have to repeat his entire green tea–collecting trip.

In the letter accompanying his first shipment, Fortune added humbly, "I shall be grateful to receive any instructions which you may think it necessary to give me . . . addressed to the care of Mssrs Dent who will forward them to me."

It had been nine months since he had received instructions from India's botanic gardeners on what seeds to collect. Fortune communicated frequently with Her Majesty's consuls in Shanghai and Hong Kong, and they had orders to be helpful to him. But there had been no news out of India House in London and nothing further from the subcontinent: no instructions, no suggestions, no acknowledgment of his task at all.

9

Calcutta Botanic Garden, March 1849

Whereas March in England saw gardeners clearing dead winter undergrowth to make way for the arrival of bulbs and perennials, March in India was alive with full tropical majesty. Nothing in the exuberant botanic garden of Calcutta resembled the timidity of an English spring; the seasons here went from hot to wet and then back again. March was still considered "the cold weather," but as one traveler wrote, "In India 'cold weather' is merely a conventional phrase and has come into use through the necessity of having some way to distinguish between weather which will melt a brass door-knob and weather which will only make it mushy." The holiday of Holi was upon India, celebrated by marauding gangs of young men who doused strangers with cold water as the year marked its turn toward pitiless summer.

Hugh Falconer was reddening in the sun as he walked through Calcutta Botanic Garden in March 1849, surveying rows of tea seedlings that his *malis*, or gardeners, were busy transplanting and pruning. A burly, barrel-chested Scotsman, Falconer was director of the garden, which was the country's de facto department of agriculture. Falconer concentrated on horticultural networking and policy making, on "improving" the agriculture of India and fixing the "unaccountable discrepancy" between the richness of the country's soil and the poor quality of its agricultural products. Economically valuable plants such as teak, tobacco, coffee, and indigo were shipped to Calcutta, the capital of colonial India, from all over the empire for distribution within India. Falconer was an East India Company man in the middle of his career, but Calcutta's tough climate had aged him prematurely. He had already been home to England once on sick leave, although he was only forty-one years old. It had not become as hot yet as it would be in the months to come, and although the monsoon would bring some relief, it was at the cost of continual deluges that would fill the city's streets, turning the sewers into rivers to the sea. It was enough to make a man like Falconer wonder whether he had chosen his career well.

But he was needed in Calcutta as he awaited a shipment of plants and seeds from China on which the company was desperately depending. The tea the *malis* were currently cultivating in the Calcutta garden was of little consequence compared to what was due to arrive any day. The Calcutta tea was good enough for experiments on planting depth and pollination but was not suitable for drinking; it came from native Assam stock, tasted bad, and was ill-suited for the high-altitude company gardens in the Himalayas.

Fortune's seeds would be placed under Falconer's care. Falconer, like most naturalists of Great Britain, was self-reliant and systematic, accustomed to working alone and to being right—much like Fortune. He was a surgeon and a dedicated East India man, and as good a custodian for Fortune's tea seeds as he could wish to have. Falconer believed tea was crucially important to the future of the company, that the garden at Calcutta was at the crux of India's tea project, and that his legacy as superintendent depended on the success of this scheme. Fortune and Falconer, two gardeners, were of one mind when it came to the need to steal tea from China.

The Calcutta Botanic Garden itself was a magnificent sight: "Trees of the rarest kinds, from Nepal and the Cape, Brazil and Penang, Java and Sumatra, are gathered together in that spot. The mahogany towers there and the Cuba palms form an avenue like the aisle of some lofty cathedral. Noble mango trees and tamarinds are dotted about the grassy lawns; and there are stately casuarinas around whose stems are trained climbing plants. There are plantains of vast size and beauty from the Malay Archipelago, and giant creepers from South America. The crimson hibiscus and scarlet passion-flower dazzle the eye, and the odour of the champak and innumerable jessamines [*sic*] float upon the breeze," said a visitor.

Located on the west bank of the Hooghly, the garden stood just opposite and around a bend in the river from Fort William, the high-walled seat of the East India Company administration in India. It was as much a park as a laboratory, a place for picnic lunches to keep the hustle and chaos of Calcutta at a civilized distance. "Every step is a surprise," acclaimed one visitor. Observed another, "The Botanic Gardens would perfectly answer to Milton's idea of Paradise, if they were on a hill instead of a dead flat." Covering three hundred acres just below the city, the company's garden, like other colonial imports to India, was noted for the "order and neatness of every part, as well as with the great collection of plants from every quarter of the globe." It was one part of Calcutta to which Kipling's epithet—"this God-forgotten city"—seemed not to apply.

The Botanic Garden dated from the early days of botanical

imperialism, around 1786, when a gardening-obsessed infantry-
man suggested to the government that a site for the study of India's flora might prove useful—and profitable—to the shareholders
of the company. Initially directed to introduce nutmeg, cinna-
mon, cloves, peppers, and breadfruit into the subcontinent, the
gardeners of India discovered that Calcutta was a poor home for
equatorial species and an even worse place to grow many valuable
trade goods.

Although tropical farming failed there, it was not a total
loss as a laboratory. Indeed, the garden became central to the
program of global plant exchange and commerce for the East
India Company. The garden "has fortunately always been a pet
with the respective governments of India; and, in consequence,
considerable outlays have, from time to time, been made, to
keep it in the most perfect order and efficiency. To enable trav-
ellers, and others, to avail themselves, as much as possible, of
the [benefits] of this establishment, the superintendent has a
supply of seeds and roots always ready for those who may apply
for them," wrote Dr. Royle, senior botanist of the East India
Company, who had originally hired Fortune for the China tea
assignment. On the banks of the Hooghly exotic specimens
were bred, cataloged, and numbered, recorded for history and
puzzled over for trading purposes. Under Falconer the Calcutta
garden was the crucial nexus for information and plant exchange among the smaller company gardens in the Indian provinces. Seeds and saplings were shared, native Indian plants
were sent to gardeners all over the world, and new ideas were
reported and discussed. India was an extremely collegial place
to do science.

The broader aim of the Calcutta garden was to connect the
natural glories of the British Empire to Kew Gardens in En-

gland. Kew was the center of botanical research for the entire world; all seeds, shrubs, specimens, and herbaria were forwarded from the empire's outposts to Kew's gardeners, the ultimate arbiters of horticulture. Practically speaking, however, Kew's centrality was secondary to botanists in the field who were busy improvising their way around the world, cataloging and describing every living thing, and trying to make new plants grow. Developing plant-based industries on foreign soil was the stated aim of company botanists, a task they accomplished with unparalleled skill. The Calcutta gardens introduced teak and mahogany for the timber trade, distributed hardy grains to feed India's famished peasants, and conquered malaria with quinine produced from the bark of the South American chinchona tree.

Falconer had come to the directorship of the gardens as the successor of Dr. John Forbes Royle, the company's senior botanist. Royle had first appointed the fledgling Falconer his deputy in exploring the Himalayas and, within two years, named him to the position of superintendent at the Saharanpur Botanic Gardens in the Eastern Himalayas.

Falconer was skilled and erudite, a botanist but also an avid if amateur paleontologist. He was the first to articulate the evolutionary theory of "punctuated equilibrium," which holds that sexually reproducing species will show long periods of stasis for most of their fossil record, but when evolution does occur, it appears to happen rapidly and all at once. While working in the Himalayas, Falconer discovered one of the first fossilized monkey skulls—a fact noted by Charles Darwin in his *On the Origin of Species by Means of Natural Selection*. On home leave to Britain in the middle of his Indian career, Falconer shipped to the British Museum an incredible five tons of fossil bones embedded in their rock matrices.

Falconer and Royle both strode in the footsteps of India's great naturalist Nathaniel Wallich, the man who might properly be called the founder of Indian tea.

For more than thirty years Wallich, a Dane, was the leading botanical authority on the subcontinent and oversaw the Calcutta garden. He "left his country young, and has devoted his life to natural history and botany in the East. His character and conversation are more than usually interesting; the first all frankness, friendliness, and ardent zeal for the service of science; the last enriched by a greater store of curious information," wrote an acquaintance. Wallich was probably one of the first Europeans to taste Indian-grown tea, although he didn't realize it at the time.

There had been ongoing debates in the early half of the nineteenth century among botanists over whether any tea existed naturally in India and, if so, what it looked like. Wallich, who joined the company in 1817, was initially a skeptic and rejected the possibility. Since he was the leading botanist in India, his opinion was definitive, and thus he nearly scuttled the success of Indian tea from the outset.

When the East India Company annexed Assam Province, next to Burma, to the rest of its British possessions in 1824, two brothers, Robert Bruce and C. A. Bruce, an ex-army businessman and tea merchant, respectively, went to the new territory looking for trading opportunities. There they found what they believed were tea plants growing wild on the hillsides. They spoke to natives, who steeped a brew from the leaves and chewed them to relax. The Bruces transplanted some seedlings to a private garden and sent samples to Wallich.

He steeped some of the dried leaves, tasted the golden brew, looked again at the accompanying sample of whole leaves of the same plant, and dismissed the lot as just another unremarkable evergreen. How could it be tea? reasoned Wallich. The area where these leaves came from was at sea level, while everyone knew that Chinese tea grew only in higher, mountainous regions. Seven years later another set of Assam samples, these from an Indian army lieutenant, were brought to Wallich's attention, and he once again refused to confirm the existence of native Indian tea plants.

But as the company's position looked increasingly insecure in the Orient, pressure mounted to find a way to grow tea elsewhere. With the East India Company monopoly in China nearing its end in 1834, the governor-general of India established a committee in Calcutta to further investigate the possibility of growing tea in the British dominions there. Wallich, a conservative, was as much a follower as an ideas man and was easily influenced by popular sentiment. With political pressure being applied by the company to find a viable way to produce tea in India, he was finally encouraged to admit that the leaves he had been sent were actually tea leaves—that, in fact, tea *was* native to India. With heavy prompting by the company and encouraged by the presence of his protégé, Falconer, Wallich finally took the risk in favor of scientific discovery.

From the fiercest of tea skeptics, Wallich became one of Indian tea's most avid champions. Together with Falconer, he took lengthy trips to explore and map Darjeeling after the company seized it as a "gift" from the rajah of Sikkim and annexed it to India. Wallich dedicated scarce ground in the Calcutta garden to tea seedlings. After collecting land data from the vast network of surgeons employed in the remotest parts of India, he came to believe wholeheartedly in the future profitability and sustainability of a tea economy on the subcontinent. He conducted a survey among company surgeons to research the most likely places to establish tea estates, and in the end chose the gardens under the care of his former pupil, Falconer: Saharanpur, in the high-altitude Himalayas, would be the site of the tea-growing experiment.

Falconer, then a young man, was a keen supporter of the tea project and may have been the first to plant tea seed in the Himalayas. In those trials the seed was of very poor quality, having been smuggled out of Canton where even indigent Chinese peasants resented such a paltry drink. Falconer persisted and ultimately produced what looked and tasted to him like a fair facsimile of the Chinese original. The question remained, however, whether Indian tea would catch on in England.

In January 1839 the news that Indian tea from Assam had arrived in London set British imaginations on fire. Connoisseurs considered it at the very least a curiosity, but it might possibly prove to be a great prize. All the major London tea merchants as well as the press were in attendance at the Mincing Lane auction where it was introduced. The event reflected a general uneasiness, even a low-level panic, about the stalemate in the Orient because it took place just prior to the First Opium War between Britain and China.

The night before the auction, the Indian tea was brought before tea inspectors, men whose noses and tongues dictated the blends and tastes of British tea drinkers. Their assessment found the Indian tea leaves to be dark and leathery, their brew bitter, and their aroma heavy. Even with those marks against it, however, the Indian tea was declared "of reasonable quality."

First on the block were "chops" (cases) of the finest quality teas. In heated bidding the first round of tea fetched record prices. The atmosphere was electric, and with each successive lot, the prices escalated as the crowd worked itself into a bidding frenzy.

Finally the hammer went down on the last lot of Indian tea, the chops of the very lowest quality, the filler and twigs of broken, damaged leaves. This last, inferior lot fetched a higher price than the earlier, superior teas, a staggering 34 shillings per pound (roughly $168).

Those in the room recognized the sale for what it was: an outbreak of tea hysteria. The desire for Indian tea, fueled by its novelty and the looming threat of open hostilities in the coming First Opium War, would not be extinguished, though it would take another twenty years to bring it about.

In 1847, Nathaniel Wallich decided to retire at the age of sixty-one. This opened up the Calcutta post for a successor, and there was no better choice than Falconer, who had been named a fellow of the Royal Society in 1845 and had earned a medal from the Geological Society of London. To add to Falconer's honors, the company named him professor of botany at the Calcutta Medical College. Although the directors of the East India Company thought that Wallich—who had nearly thirty years of service to the company—was grossly overpaid, the Honourable Board elected to continue his inflated salary for Falconer, so important was the post of horticulturalist to India.

It is likely that the Himalayan tea experiments were on Falconer's mind in March 1849 as he made his way back from the experimental fields in Calcutta to the caretaker's cottage in the garden—past native orchids in full bloom, past the artificial mound that his mentor had hastily constructed so as to show off some trees he had collected in the Himalayas, and past a great and ancient banyan with almost two hundred prop roots spanning nearly an acre. Falconer walked past the peaceful ornamental lakes toward the corpse-strewn banks of the Hooghly beyond. The breeze reeked, as it always did, of sewage.

He knew that this season's tea planting would be another disappointment, for there were still not enough seeds available for it to be truly profitable. Falconer must have privately wondered if Fortune was on a fool's errand; transporting seeds from the most far-flung provinces of China might well be beyond the capacity of even the most skilled botanist. But it was more than just the fate of Fortune's mission that vexed Falconer; it was also the personnel problem in the Himalayas. He was already having doubts about the man who had taken over from him in Saharanpur, a young botanist named William Jameson. Even if the Himalayas received enough imported seeds and saplings, as well as Chinese manufacturers to teach tea making and packing skills, it seemed increasingly unlikely to Falconer that Jameson would be up to the task of managing the project properly.

10

Saharanpur, North-West Provinces, June 1849

Calcutta, the seat of British rule in India, was orderly and tame by comparison to India's remoter regions. Whereas Calcutta's Botanic Garden was well cultivated and manicured and generally civilized, the Himalayan gardens of Saharanpur were lush and wild and founded on a site that had formerly been an old Rohilla garden. (The Rohillas were Pashtun invaders from Afghanistan who had once commanded Northern India and built themselves luxurious pleasure palaces in the hills.)

The region was temperate and hilly and had ample rainfall, and every living thing seemed to thrive in the nourishing Himalayan soil. Tigers, panthers, and lynxes roamed the rhododendron forests like creatures from fairy tales. Celebrated Rajput warriors, dressed in red silks and wearing handlebar mustaches, bred magnificent horses in the mountains near Saharanpur. Because Saharanpur was located on the border between the Persian and Asian horticultural zones, plants from both regions flourished there. Although it was a horticultural paradise, gardening there was not always easy on a white man. "A *mens sana in corpore sano* [healthy mind in a healthy body] is absolutely necessary to resist this dreadful climate: the work is very hard, the sun a terrible enemy; there are many comforts wanting, scarcely any society, and in his daily habits a man has to exercise an enormous amount of self-denial and discretion if he wishes to retain good health," wrote a tea planter. To the Britons committed to building a tea industry there, Saharanpur was remote, its heat oppressive, and the living conditions primitive. "To these discomforts add one more—an unquenchable thirst that is ever present, but is particularly noticeable after severe exertion, when the desire to drink . . . is painful to a degree. This insatiable thirst is the great curse of the climate, and has accounted for many good men who have gone under the *matti* [earth]." These hardships notwithstanding, the large tracts of land and rainy climate should have made Saharanpur the ideal home for the first transplanted Chinese tea.

But first the plants would have to get there in something like the condition in which Fortune had shipped them. As it turned out, neither they nor the seeds had an easy trip.

Having personally accompanied the seedlings to Hong Kong, Fortune sent them seaward. But luck was not on his side. Although the shipment was intended for Calcutta, for some reason it was diverted to Ceylon (modern-day Sri Lanka). Fortune's shipment could not have been the primary cargo: ten thousand tea seeds filled only five gallons and took up at most five crates when packed separately in bags of sand. Fortune's thirteen thousand seedlings were in glass cases, but the fate of a few botanic specimens were likely of little concern to a captain who had more pressing priorities. It is entirely possible the ship was delayed by bad weather or made an extra stop to offload higher-paying cargo. Those were the glory days of shipping under sail, and merchantmen frequently made unplanned dockings for any number of reasons—repair, renegotiation of contracts, barter, or bad planning.

After completing whatever business took it to Ceylon, the ship reversed course and made its way east again toward the port of Calcutta. Upon arrival it fell to Hugh Falconer to assume or delegate responsibility for Fortune's cargo and then to transport it to its ultimate destination in the Himalayas. He collected the shipment, signing for it just after the spring holiday of Holi on March 23—two full months after the plants had set sail from China. (In two months a fully rigged tea clipper could sail halfway from Hong Kong to London around the Horn of Africa.) Even with the unexpected delay, Ward's cases should have guaranteed that the plants would still be alive and healthy as long as the cases were kept in sunlight and sealed from salt spray.

Upon receiving the shipment, Falconer was careful to do nothing to disturb the cases, much as he might have liked to open one and investigate its contents. He contented himself with peering into the terraria. As far as Falconer could tell, all looked to be in order; the plants would take care of themselves. He left the cases outdoors, shaded from the worst of the sun.

Within a few days of their reaching Calcutta, Falconer ordered the cases to be loaded onto a steamer that was traveling up the Ganges bound for Allahabad, halfway up-country from Calcutta to the North-West Provinces. It was there, apparently, on April 12 that an official at the highest levels of local government, whose name has since been forgotten, did the very worst thing he could have: He opened the cases. It was certainly an understandable impulse. Motivated by a desire to ascertain the condition of the precious cargo, the official, or one of his underlings, broke the seals. He even reported to his superiors that the plants inside were doing well.

The transfer to another vessel in Allahabad took longer than expected. There was a drought that year, so the Ganges was low, which meant the seedlings could not be shipped by steamer up-river to the company gardens in the Himalayas until the rains came with the first summer monsoon, still six weeks away. Nonetheless, the tea plants, according to a report, were tended to by gardeners and in good condition.

The last leg of the long trip up the mountains to the Saharan-pur experimental tea gardens was made first by steamer and then by oxcart. In the Himalayas, Fortune's tea would be met by William Jameson of the company's botanical outpost there.

High in the mountains Fortune's newly arrived tea was received and inspected. The results were disastrous. Only one thousand of the thirteen thousand seedlings remained alive, and the survivors were covered with fungus and mold. The glass cases had the stale reek of rot. The success rate of Fortune's first shipment of tea stood at a pathetic 7 percent. Gamely, Jameson did what he could to rescue the plants, picking out and discarding the dead ones, nursing the barely living ones, and ordering the Saharanpur *malis* to replant the few robust specimens in the fertile mountain soil. Even after these steps had been taken, however, the survival rate of Fortune's first shipment dropped to a mere 3 percent.

The fate of the seeds was even more disheartening. "The result has been an entire failure, not one seed having germinated. I lately removed a number of seeds from the beds to ascertain their condition and invariably found them to be rotten," Jameson wrote.

Despite spending an entire year of planning, collecting, packing, and hoping, Fortune had done nothing whatsoever to advance the cause of tea in India. He had made no contribution to the nursery stock of the Himalayan tea experiment. What should have been a triumph, the culmination of his first year of tea hunting, was a catastrophe of wasted energy and worthless tea stock.

Behind the gentlemanly façade of the East India Company there were often hidden rivalries and simmering tensions, and within the Indian network of company gardeners this friction often centered on the cultivation of tea. The "smaller" gardens on the subcontinent were only so in relative terms: By square mileage, the experimental tea gardens in the Himalayas occupied more ground than was available to the Calcutta Botanic Garden. But the Calcutta garden bore the imprimatur of colonial authority, and in the hierarchy of the company, authority must always be respected. As the superintendent of Calcutta, Falconer issued his orders, and the provincial gardeners were expected to follow them. William Jameson, superintendent at Saharanpur, was one such gardener.

Both Falconer and Jameson were Scottish; both were surgeons and naturalists; both had studied under the finest scientific minds of Scotland; both were in the employ of the company on the subcontinent; both took a keen, almost proprietary interest in the fate of the tea experiment. But there the similarities ended.

Young Jameson was a great deal less professionally polished than Falconer and something of a bumbler. While other naturalists in the medical corps were mapping uncharted territories and unraveling natural history puzzles, Jameson found himself inadvertently imprisoned in Peshawar for trespassing. He did not rise to prominence on the broad sweep of his learning or on his imagination but through a dogged navigation of colonial hierarchy and, most likely, thanks to a nepotistic thumb on the scale.

Jameson's uncle, Robert Jameson, was a celebrated professor of geology and an expert on India, a peer to Wallich and Falconer and the teacher of Charles Darwin. William Jameson was sufficiently politically savvy to ride his uncle's coattails.

Jameson was prone to answering every official letter at interminable length. He issued elaborate protocols and detailed orders for matters that had never before needed such painstaking attention. The skills that may have enabled him to thrive in a bureaucracy were hardly transferable to botany, however. Jameson wrote pamphlets in which he held forth on the theory of garden design, the state of the weather, the political situation in China (where he had never been), and the preferred methods for planting each and every species under cultivation. Although exhaustive, his recommendations were seldom followed by his superiors and often completely ignored.

Jameson ran the Himalayan tea gardens as if they were a factory, concerning himself mostly with the management of men and resources. His letters to the government are as thoroughly worked out as any contemporary business plan, full of details about size and scale, and what the profit per acre could be if he only had the manpower and assets.

The government was eventually compelled to chide Jameson, asking him to be more measured in his enthusiasms: "As it is evidently the wish of the Honourable the Court of Directors that the experiment should be conducted on the most liberal scale the Lieutenant Governor is pleased to sanction your proposal as to the extent of operations to be ultimately reached. It is evident however that some time must elapse before you can employ to advantage so large an establishment as you propose. You will be careful not to extend your establishment till you have the means of fully and profitably employing them."

Although management and business planning were valuable skills, they tended to distract Jameson from his botanical studies. Plants grew poorly under his care, and whatever zeal he had for the subject was continually undermined by his lack of botanical knowledge and failure to investigate any given problem thoroughly. In his eager rush to document everything and gain approval from his superiors, Jameson erred in his science. He had been employed as a chief gardener even though he had devoted himself to studying zoology and geology. Reading Jameson's ponderous declarations on the anatomy of tea is a little like listening to a parish organist's recital of Beethoven—the notes are in the right place, but the music sounds wrong. The natural beauty of the piece is lost.

Jameson's scientific mistakes were many, costly, and easily avoidable. For instance, he accepted as holy writ that the tea plants were of two different species even though the relevant studies had been done within the confines of a laboratory in London by people who had never studied the plant in situ and Fortune, meanwhile, had published his findings contradicting the matter. Although there were tea plants to experiment on in his very own Himalayan garden, Jameson accepted out-of-date hearsay and received wisdom without ever doing any further study of his own. He accordingly designed the vast company gardens around a single, easily correctable misconception.

It was not simply that Jameson was often wrong but that even in the face of a self-evident truth he stubbornly continued to hold the opposite position. Although the Chinese workers in his gardens told him that green and black tea were the products of the same plant, he disregarded their greater knowledge. Even worse, he was actually killing the tea plants. In the Himalayan gardens he set aside flat lands for tea and used flood irrigation on them—a system he used for years despite the chronically sickly appearance of the overwatered plants in the swampy fields. Had he used his power of observation, Jameson might have deduced that tea would in fact do better in rugged, sloping terrain. The roots of *Camellia sinensis* need good drainage; otherwise the bushes become waterlogged and moldy and can't produce healthy young buds. Fortune had already published work on tea cultivation in China after his first trip, but Jameson seems to have paid no heed whatsoever to the science of his day.

Most damagingly, Jameson was not aware of the science that was the basis for the Wardian case. "They have received water all the way from Allahabad and had it not been for this circumstance there would not have been so many plants alive," wrote Jameson to his superiors, knowing—as all company men did—that his assessment would find its way to Falconer, who was predictably outraged. (All communications were public unless explicitly indicated to be off the record, and even then the Revenue Department was notoriously careless with information.)

It was not entirely Jameson's ineptitude—or that of the men who followed his instructions—that had decimated the consignment. "Many of the panes of glass in the cases, too, were broken," he recorded and, ever the dutiful company man, naturally issued a dispatch on the matter.

> In future when cases are sent from China, the following instructions ought to be attended to:
> A careful Mallee ought to accompany the cases [from Calcutta] and before leaving the Botanical Gar-

dens he ought to be provided with a screwdriver and taught how to unscrew the frames of the cases so as to enable him to water the plants occasionally that in every second day and again rescrew the frames. . . .

On arrival of the cases at Allahabad they ought to be dispatched to Saharamfore [*sic*] in a government wagon appropriated to the purpose and well covered to protect them from the heat of the sun and under charge of a careful person.

If anyone had ever bothered to follow them, Jameson's instructions for the boxes would have guaranteed that all plants would be dead on arrival.

He likewise offered new and eccentric instructions for the care of future seed shipments:

With reference to seeds. On receipt of the seeds in Calcutta parcels or boxes ought immediately to be sent to the Botanical Garden to the Superintendent who ought to receive instructions to open them and to inspect them and forward one half in parcels by letter to Saharamfore, the other half might be sown in flower pots or cases and kept at the Garden until they germinate and the growing plants then forwarded under charge of a careful person by the steamer to Allahabad.

Jameson declared there was a historical precedent for this. The only other China tea in India, those bushes grown out of Canton seeds from an earlier shipment, had come to the subcontinent in the same fashion.

By sowing seeds on receiving them from China, from Doctor Gordon, Dr Wallich was entailed to supply the [Himalayan] plantations with the first tea plants and from these and their produce the plantations now thriving have been formed. By adopting this plan there would be two chances in form of a successful issue as if the seeds did not germinate in the plantations owing to being so long out of the ground, they might do so in Calcutta.

It seems likely that Fortune's seeds were probably doomed by the time they arrived in India and that Jameson's suggestions were of no consequence either way. Planting seeds immediately on receipt in Calcutta was a reasonable theory on how to save them, based on past evidence, but probably incorrect. Gordon's earliest shipment of seeds had been picked in Canton, so the time between harvesting and sailing was negligible—a matter of weeks or a few months at most. Fortune's seeds, however, had been picked on the Wangs' land in Xiuning County; traveled months by river and canal boat to Shanghai; were repacked there over the course of several weeks; were shipped to Hong Kong and unloaded there and then reloaded for India; were diverted to Ceylon and thence to Calcutta—all over a period of at least six months. If the seeds were to thrive in India, Fortune would need to find a better method of shipment than simply packing them in bags of sand and hoping for the best. From now on seeds would have to be treated more scientifically.

Confident of his own understanding of what had gone wrong, Jameson played the system, putting the blame for the failure on someone else. Clearly Falconer could not have inspected the cases on their arrival in Calcutta, nor had he adequately provided for their safe dispatch to Allahabad. It was Falconer's poor garden-ing skills that were to blame, according to Jameson, since "I may here remark that the arrangements made by the Commissioners of Allahabad for the cases received were of such a nature as to have caused entire success had the plants reached that place in anything like good order."

Falconer simply responded by forwarding Jameson a scientific article on Wardian cases and how they worked. He copied it to Jameson's superiors in the North-West Provinces, to the Revenue Department of India at Calcutta, and to Royle at East India House in London. Even Fortune received forwarded copies of the Jameson/Falconer correspondence. Practically every man in the chain of command bore witness to the conflict between the two rival gardeners in India.

II

Ningbo to Bohea, the Great Tea Road, May and June 1849

In May the riverbanks of China were aflame with new growth. Buds ripened on the boughs—apple, cherry, and hawthorn blossoms heralding the awakening of the natural world to the sun. With the arrival of spring Fortune had a major new project under way. He had hired another small junk to sail out of the coastal city of Ningbo, a lesser treaty port, for the black tea hills of Fujian. Now he was nearing his ultimate prize.

He stood at the prow watching coolies working quayside in the warm salt breeze. He was again journeying deep into China. From the mouth of the Yangtze he would travel southwest, headed for the fabled Wuyi Mountains, source of the finest black teas. Although black tea and green tea were products of the same plant, as he had established, the two varieties were never grown in the same place. Fortune sought to obtain black tea stock from its most celebrated region.

He was concerned that there were no reliable reports on which route would take him there in time for the second flush of tea picking: "I felt rather low-spirited; I could not conceal from my mind that the journey I had undertaken was a long one and perhaps full of danger. My road lay through countries almost unknown. . . . But the die was cast, and, committing myself to the care of Him who can preserve us alike in all places, I resolved to encounter the difficulties and dangers of the road with a good heart," he recorded.

His fretful winter pondering the fortunes of his green tea consignment was behind him. He still had no news as to the fate of the Ward's cases and would have none for at least another season. As far as he knew, his green tea plants had arrived safely in the Himalayas, and all was well.

The trip to the black tea mountains was Fortune's most daring yet. He was going deeper into China than any Westerner had ever gone, as far as he knew, traveling over treacherous terrain. He planned to journey for three months by boat, sedan chair, and foot—a trip of over two hundred miles, most of it by land, all of it uncharted, and almost entirely uphill. In the Bohea hills, also known as the Wuyi Mountains, he was seeking the tea most suited to British tastes—the blackest, gentlest oolongs. (The growing demand for black tea in England was due in no small part to a glut of sugar from the West Indies and the Caribbean. Black tea takes sugar; green tea does not. When Fortune discovered that green tea was poisoned with dye, British tastes turned even more toward black tea.)

There were no black tea gardens as of yet in India, only green. No one there really knew how to manufacture black tea, and in any case the transplanted Chinese gardeners did not have access to black tea stock. If Fortune had returned to Europe without introducing black tea plants from the best black tea districts, he would have neglected his full mandate from the East India Company.

Wang and the coolie would not be accompanying Fortune on this trip. Initially he had thought about sending them on the mission to Bohea without him, daunted by the idea of traveling through unmapped parts of rural China at a time when rebellions were sweeping the countryside. Fortune was confident that they could deputize for him; Wang and the coolie were reasonably well trained as plant hunters and knew what Fortune was seeking when he went prospecting. But he had reached the end of his tether on the previous trip, worn down by their constant conniving and scheming. He could not bring himself to abandon the hunt, nor did he feel he could entirely trust them. He would have no way of knowing whether his servants had indeed gone all the way to the Wuyi Mountains to collect the plants or whether they had stopped short at inferior black tea gardens. He also assumed that when Wang and the coolie weren't plant hunting, they were likely to be squandering their time—and the company's money— on a prolonged inland vacation. Fortune, ever the enterprising traveler, was going himself. "There may also have been a lingering desire to cross the Bohea hills and to visit the far-famed Wooe-Shan," he wrote.

Wang, however, was not without his uses. Fortune still considered the trip to Sung Lo and Wang's family land an unqualified success. For backup and additional reassurance, though, he had decided that collecting a second batch of green tea seeds, ready for the following year's planting, would assuage the last of his qualms. The cost of doing so was negligible, after all—only the price of Wang's time and travels—so he ordered the translator to return home and resume collecting green tea seeds.

Fortune now required a new servant to accompany him on the long journey, so he hired, after many inquiries through Dent's comprador, a knowledgeable body servant, a portly man named Sing Hoo. With his proud and dignified bearing, Sing Hoo was "powerful and spirited" and had once been in the service of a high-ranking mandarin affiliated with the imperial family at Peking. His elevated status was clearly visible in his straight shoulders and in the proud stretch of his neck. Hoo came equipped with the insignia of his former office, a small triangular flag that bore the arms of the Imperial Court, which, he said, was a gift from his former master and amounted to a sort of passe-partout throughout the country. To anyone owing fealty to the emperor, this flag signaled that the travelers were under the court's protection. "I confess I was rather sceptical as to the power of this flag, but allowed him to have his own way," Fortune recalled. The servant carried the flag everywhere, rolled up and at the ready.

Sing Hoo was from Fujian Province, where the Wuyi Mountains were situated, which meant he spoke Fukienese, the local dialect. Although China had a standard and official court language, each province had its own dialect that was almost unintelligible to outsiders. Fortune's halting Chinese would most likely have been a pidgin version of Shanghaihua—the formal tongue of Shanghai—for he would have learned it from the men he hired there. He might have spoken a touch of Cantonese, or Guandonghua, also, like so many of the British merchants and government men who were posted in China, but mostly he communicated in pidgin. In the mountainous areas of remote Fujian he would have been hopelessly lost linguistically. Sing Hoo was his new mouthpiece.

Fortune's previous fear of exposure as an outsider was allayed on this journey. The worrying skirmishes on his earlier green tea trips had been instigated in large part by his servants; his Chinese was awkward but by now passable; he was competent with chopsticks; and his clothing marked him as a man of the realm. More confident now, Fortune thought it entirely possible that no one would detect him at all—he was going so far from the coast that it was probable no one in the region had ever seen an Occidental face.

Not that this journey was without its dangers. Besides the rumors of peasant rebellions, Fortune was as yet uncertain about Sing Hoo: "The guide I had with me was not fully to be depended upon." But he was at the very least a welcome change from Wang and the coolie's outright hooliganism. If officials were being slaughtered and the poor were seeking retribution for their suffering, as they were said to be in the hills of Fujian, Fortune's silken costume identifying him as a ranking mandarin would scarcely provide much protection. Command of the local tongue and the triangular flag of the Imperial Court might well provide all the help he would get.

After imperiously ordering the men on the junk to store his master's belongings correctly, Sing Hoo informed Fortune that it was time for the journey to begin; Fortune must become a different "outward man" and clothe himself in full Chinese dress. Fortune's braid still hung down his back. He removed his Western clothing—his hard-soled shoes and buttoned-up jacket—and donned the wide, flowing garments of a Chinese official.

"I doubt whether my nearest friends would have known me," he wrote. "I scarcely recognised myself."

"You will do very well," Sing Hoo told him.

Fortune, Sing Hoo, and all their gear floated lazily upriver, the boatmen calling to one another as they poled around the bends. They sailed past walled cities whose ramparts dated back thousands of years to the times of plague and savagery in Middle Europe. (In China those same times saw the rise of highly advanced civilizations, social organization, and achievement.) They traveled along an ancient network of canals that stretched across

China like a spider's web, connecting the vast area of the Center Country with everywhere else of importance.

Within a few days the men came upon a traffic jam in the canal, a bottleneck at a junction point where fifty junks were floating helplessly alongside one another. Fortune's junk queued for its turn on a windlass that would raise it onto a ramp leading to a higher canal. Stevedores on the shore were fixing ropes from the windlass to each boat's prow, winching it up inch by inch. The delay was only an hour or so—not overly long, given the average waiting time in China—and all the boatmen on the river, who were never in a hurry to begin with, spent the time sunbathing and spitting, playing mah-jong, and enjoying their time off and the spring sun.

All except for one fellow, that is, who sailed astern of the queue. He huffed and spat and swore, growing increasingly impatient. The angry boatman maneuvered his boat toward Fortune's junk, knocking others aside, shouting, and threatening any captain who would not get out of the way. Most of the boatmen ignored him, allowing the interloper to have his way, but when he reached Fortune's craft, the junk's captain shouted, "You cannot pass this boat," perhaps more in an attempt to placate Fortune than to provoke a fight. He jammed the nose of the junk against the canal wall, closing the gap so that the irate sailor could not pass.

Sing Hoo now also stepped into the fray, determined not to let any man outface his master.

"Oh, but I will pass you!" the obstreperous captain rudely insisted.

"Do you know," Sing Hoo shouted back, "that there is a mandarin in this boat? You had better take care what you are about!"

"I don't care for mandarins," shouted the angry captain, spitting out a sentiment shared by many but spoken by few. "I must get on."

"Oh, very well," Sing Hoo replied calmly, "we shall see." The servant went down into the cabin, took his talisman from his luggage, and unfolded the triangular yellow flag. He walked out again into the sunshine, smiling to himself, and hoisted the banner on the mast.

"There," taunted Hoo. "Will you pass us now?"

Fortune was dumbfounded when the captain stood down, apologizing profusely and becoming "all at once as meek as a lamb." From then on he did nothing but sit quietly on the stern of his boat, eyes cast down and waiting his turn like everyone else in the canal.

Fortune smiled; perhaps he would find safety in this countryside after all. Perhaps the path ahead would be strewn with more luck than he had imagined. Sing Hoo's flag had marked him as an important mandarin, a *lau ye*, a "sir."

Fortune penetrated deeper into China, sailing from town to town in Zhejiang and then journeying on by sedan chair and mountain footpath through Jiangxi Province. As he passed through ruined villages, beggars reached up to him, holding out skeletal hands for a few pieces of copper cash or anything that might be useful in barter. Fortune was touched by the supplicants' poverty but horrified by their scabrous faces and missing limbs. It did not escape his notice that the peasants of China were suffering blow after blow, season upon season, so many bitter lessons at the hands of their vengeful or indifferent gods.

The trip to black tea country was a slow climb that would take almost three months to complete. It was a beautiful time of year:

"The lowlands were now much broader—the hills appeared to fall back, and a beautiful rich valley was disclosed to view. Groups of pine-trees were observed scattered over the country. They marked the last resting places of the dead, and had a pleasing and pretty effect." The Chinese would often plant trees over graves as a sign of respect for their ancestors. As the rich rolling hills of the Yangtze valley gave way to China's coastal mountains, Fortune's mission took him to areas as striking as they were unnerving.

At first glance the vistas were lovely, dotted with swallows on the wing, but on closer examination China's landscape was unforgiving. The luxuriant vegetation swarmed over ancient buildings, forcing its way through dilapidated shacks. The same climate that made roses burst into flower in early summer was merciless on man-made structures. Each peasant hut looked tenuous and feeble on its patch of land, as if ready to be reclaimed at any minute by the earth. The region was lashed by high winds and was hit hard that year by flooding. What had once been among China's richest areas had been ravaged by a series of natural disasters, and famine was forcing entire clans off their land. Extreme weather had always limited the amount of arable land available to the growing Chinese nation, but the relative peace of the Qing reign meant that the population had doubled in size in the previous century. In 1849 there were more Chinese than ever farming fewer fields, while successive rains and droughts only undermined their efforts.

From his palanquin Fortune saw tea coolies carrying crates strung on poles across their shoulders as they marched in line down a mountain path, like colonies of ants on the move. The tea caravans and the profusion of beggars in the area did not seem to have any relation to each other, yet the beggars had gathered there precisely because the tea route, like the fabled silk route, was one of the most valuable avenues of trade in imperial China.

Tea commerce was then worth almost $26 million a year in revenues (equivalent to about $650 million today).

It was an uneasy time to be so far off the beaten track. The blow inflicted by the First Opium War on the people of China was now being felt inland. The nation's humiliation at the hands of the West led to inflation as the beleaguered peasants paid taxes in a currency radically devalued by war debts. The foreign powers insisted on favorable trading terms, chipping away at whatever competitive advantages the working peasantry of China had ever had for their goods and labors, especially for their tea. There was palpable anger against corrupt officials who provided, at best, perfunctory aid to the suffering agricultural regions, and against the government in Peking, made up of foreign Manchus. When he could not repay a debt, one of Fortune's boatmen had his sail repossessed, which meant he could not proceed upriver and had to dispatch all his passengers and cargo at a loss. The boatman, in despair, threatened to drown himself.

If China was ripe for foreign incursion and a fertile hunting ground for botanist spies, it was equally a nation primed for threats from within. Unknown to Fortune, an insurgency was fomenting along his route. A charismatic rebel leader in the south, Hong Xiuquan, had seized the imaginations of China's impoverished and bedraggled peasantry. As a young provincial in Canton, Hong had tried to gain acceptance into the higher ranks of intellectual bureaucrats by training for imperial service, but like so many rural Chinese, he had failed to pass the examinations. His impoverished family had sacrificed everything to pay for his education, and he had tried three times to qualify for the degree that would give him the social status of wearing scholar's robes and receiving a lifetime stipend from the state. Hong was a

Hakka, a member of an outsider race, one of the hundreds of ethnic minorities in China who did not enjoy the full social standing of the Han Chinese. The Hakka were rural farmers, their women did not follow the approved practice of foot binding, and centuries after settling in southern China, they were still viewed by locals as guests in the area. The magnetic Hong took all the disappointment and rage of his failure and modest beginnings and transformed them into a creation myth of biblical proportions.

After Hong failed the exams for the third time, he grew ill, weak, and fevered, descending into a delirium full of dark and wonderful visions. In his hallucinatory state he saw a dragon, a cock, and a tiger, Chinese symbols of power, aggression, and luck. He saw demons as well and the king of Hell.

A woman greeted him in Hell and called him "son." She bathed and soothed him, wiped his brow, and cradled him to her bosom. An old man with golden hair and beard and wearing a black dragon robe identified himself as Hong's father and commanded the young scholar to change the world. He handed Hong a sword and a golden seal and instructed him to cast out and destroy all the evil demon devils.

In the vision Hong saw another man, younger and glowing, who he learned was his older brother. While Hong would gladly have stayed with his family, his older brother was impatient that Hong should return to the world. Without his help, how would the men of earth be transformed?

"Fear not and act bravely," his dream-father said. "In times of trouble I will be your protector, whether they assail you from the left side or the right. What need you fear?"

Hong awoke a changed man, although it wasn't until he happened to pick up and read a Chinese translation of the Bible, which had been disseminated by missionaries, that he was able to make sense of his dream: It had been the Christian God who had spoken to him. The "older brother" in the dream was none other than Jesus Christ, which meant that he, Hong Xiuquan, a poor Chinese peasant, had to be the younger brother of Christ and, like him, a son of the one true God.

Hong shared this revelation with his neighbors and began to preach a fire-and-brimstone version of Old Testament Christianity. He baptized converts, called for a Christian community based on faith in God the Father, and demanded the destruction of the Confucian state and the demolition of ancestral shrines. He forbade household idols and ordered the elimination of ancestor worship. His followers were called on to shun opium, alcohol, foot binding, and prostitution.

Hong declared himself the Heavenly King of the Heavenly Kingdom of Great Peace. He raised a vast army of the Heavenly Kingdom, the Taiping Tianguo, foot soldiers to aid in God's fight against the Qing emperor's mandarins. His followers sold their property and earthly possessions and pooled their resources in a common treasury in order to bring the son of the one supreme God to rule over China. The Taiping religion was entirely new to China, a sea change from the passivity of Taoism, the conservatism of Confucianism, and the otherworldliness of Buddhist philosophies. The Taiping Heavenly Kingdom was a radical call to arms. In the south, secret societies joined forces with the Taiping in hopes of bringing down Manchu rule. The Taiping wore their hair long in defiance of the Manchu tonsure. Their army had grown to ten thousand and would ultimately number some thirty thousand members, who would colonize much of eastern China. The Taiping rebellion would sweep through sixteen provinces and destroy over six hundred cities and twenty million Chinese people within three years. Unaware that he was heading into danger, Fortune made his way higher into the mountains, crossing the path of the oncoming Taiping army.

Despite the political unrest, the small towns seemed remarkably beautiful to him and were some of "the prettiest Chinese towns which I have seen . . . an English place more than a Chinese one." His descriptions of the scenery became positively and uncharacteristically florid. The vales were "even more beautiful . . . surrounded by hills, dotted over with clumps of pine, cypress, and camphor trees, traversed by a branching and winding river, and extremely fertile. . . . The whole valley seems, as it were, one vast and beautiful garden, surrounded and apparently hemmed in by hills."

In his sedan chair Fortune was borne up and around on the narrow gravel mountain tracks, every switchback leaving him teetering between the bare rock and the sky. The road was an almost superhuman undertaking, a stairway hand-carved into a mountain face. His bearers carried him straight up the high rock karsts; there seemed to be no downward slope at all, no lulls, no respites on the trail. It was unlike any other place that Fortune had known. "In some places the height was so great that it made me giddy to look down." The valleys below were gray pools of mist. Every quarter of a mile or so the traveling group came upon a teahouse. To give his chair coolies a break, Fortune frequently stopped to sample the wares of the tea merchants, enjoying a cup of "pure bohea on its native mountains" and feeling all the more Chinese.

We find tea one of the necessaries of life in the
strictest sense of the word. A Chinese never drinks
cold water, which he abhors, and considers unhealthy.

Tea is his favourite beverage from morning until night; not what we call tea, mixed with milk and sugar, but the essence of the herb itself drawn out in pure water. One acquainted with the habits of this people can scarcely conceive the idea of the Chinese Empire existing, were it deprived of the tea plant; and I am sure that the extensive use of this beverage adds much to the health and comfort of the great body of the people.

While Fortune enthused over the Chinese way of drinking tea, he was less willing to acclimatize himself to the dubious comforts of the Chinese roadside inns where he took up nightly residence. They were dark and tiny, not much more than human stables; their walls were blackened with soot and grease from the kitchen fires. Yet the trip was a pleasure to him, for by now he was so entranced with China that he took even the poor accommodations in good humor: "I never expected to find my way strewed with luxuries," he drily observed.

As it had in the past, though, Fortune's grand plan was complicated by the personal aims and scheming of his hired help. Despite Sing Hoo's relatively elevated background, he, too, turned out to have an eye for opportunity. Fortune had intentionally tried to keep his luggage to a minimum—a few necessary items of clothing, and a sleeping mat—in order to leave more room for tea plants and seeds. Sing Hoo, on the other hand, had "a strange propensity of accumulating" unnecessary things.

"Everything good comes from Nanche," he insisted while purchasing heaps and bolts of grass cloth. This household staple, used to cover dirt floors in peasants' homes, was a few cents

cheaper inland than on the coasts, giving Sing Hoo an unexpected business windfall—one that outraged Fortune, who funded their transport.

"You see," Sing Hoo explained while conceding it was necessary to hire a coolie to carry their baggage, "we have reduced it so much that he will not have half a load. Now the carriage of this cloth will not add anything to the expenses, and the man's load will be properly balanced."

Fortune remained unmoved.

"Travelers in my country who have a goodly portion of luggage are always considered more respectable than those who have little," Sing Hoo continued authoritatively.

His continual jockeying for self-advancement made Fortune feel badly used. He would never be at ease with the Chinese penchant for personal gain; it ran contrary to his own sense of pecuniary propriety. Much to the delight of his chair coolies, Fortune often insisted on running ahead on the trail, to pick fresh seedlings, take samples of topsoil, and explore the mysteries of the ravines of the heights.

Bohea, covered with oak and bamboo, thistle and pine, was indeed a botanist's dream. As Fortune wandered, Sing Hoo followed with the cases and trowels, digging up a sample of every undiscovered species and many new varieties of familiar ones. Fortune's empty sedan chair soon became filled with clippings and cuttings from the hillside, but since these were far heavier than grass cloth, the bearers began to object; they were unable to make sense of their growing burden of "what they considered weeds." From time to time a coolie would rebel, put down his load, and deliver an angry rant about the weight of the sod he carried and how unnecessary it was.

Any sympathy Fortune had for the complaining men was

tempered by his enthusiasm for the hunt. He threatened the coolies and made promises to them, flattered and bullied them, bribed and intimidated, and offered bonuses for the increasing weight of his Wardian cases. He was resolute; the samples and herbaria were to travel with him all the way up Bohea's mountains, back down to Shanghai, and on to Kew. It was only by "dint of determination and perseverance" that his plant haul was carried several hundred miles into the tea hills and out again in complete safety. These would be the first plants ever brought from China's Wuyi Mountains to Europe.

After weeks of climbing, they approached the summits of the Bohea range, the high mountain pass that separated the interior Jiangxi Province from coastal Fujian. "Never in my life had I seen such a view as this, so grand, so sublime. High ranges of mountains were towering on my right and on my left." Fortune was overcome as he ascended ever higher into the clouds and the bamboo forests at the pass. Streams trickled down the mountainsides, and waterfalls splashed into them, joining below to become the Min River that flowed into the sea at Fuzhou where the pirates made port.

The traveling band had arrived at the "gates" of Bohea, the limestone pillars to either side of the mountain pass. These karsts were forged by nature and then worn down by water over the course of millennia. Fortune stood at the gateway to the celestial tea country, beholding "one of the grandest sights" he had ever seen and allowing himself to muse for a few moments on its beauty.

> For some time past I had been, as it were, amongst
> a sea of mountains, but now the far famed Bohea
> ranges lay before me in all their grandeur, with their

tops piercing through the lower clouds, and showing themselves far above them. They seemed to be broken up into thousands of fragments, some of which had most remarkable and striking outlines.

Fortune was among the first foreigners to try to describe the grandeur of the Wuyi Mountains. For centuries learned Confucian scholars carved poems into the base of the hills, testaments to tea, the power of the Eternal, and the might of Nature. Wuyi Shan had good *feng shui*: good wind and good water. "At a distance they seemed as if they were the impress of some giant hand." Fortune believed they were initially created by water "oozing" out of the porous rocks, forming natural carvings that were then augmented by "Emperors and other great men."

They were now at the heart of black tea country, and tea farms striped every mountainside. The weather was fine, if chilly, with sunlight gleaming off the eastern faces of the karsts, tinting them gold, while their shaded sides were "gloomy and frowning." Fortune's mind began to wander. "Strange rocks, like gigantic statues of men or various animals, appeared to crown the heights."

"Look, that is Woo-e-Shan!" Sing Hoo exclaimed.

Fortune recalled the sight of those hills with reverence: "Here I could willingly have remained until night had shut out the scene from my view."

12

Bohea, July 1849

The day had grown bright and hot as Fortune's exhausted chair coolies spiraled their way into the hills, following the narrow footpath to its steep, straight terminus in the sky.

"It is impossible to go any farther!" they insisted, so Fortune got out of the chair and walked ahead alone, climbing on his own for hours.

"Look!" they cried, forgetting their heavy loads and the rough trip, transported for a moment by the beauty of the hills. "Have you anything in your country to be compared with it?"

Indeed, Fortune had not. No matter how hot, uncomfortable, and far from home he was, or how unending the climb seemed, there was no hill or vale in the British Isles that could stand against the might and grandeur of the Wuyi Mountains. Because Wuyi's tea leaves appear to bloom a purple shade and then turn red as they ripen, Wuyi tea was known in the local dialect as *bo he*, meaning red tea, which when anglicized became bohea.

As Fortune approached the center of black tea production, he observed that the plantations looked like "a little shrubbery of evergreens. As the traveller threads his way amongst the rocky scenery . . . he is continually coming upon these plantations, which are dotted upon the sides of all the hills. The leaves are of a rich dark green, and afford a pleasing contrast to the strange and often barren scenery which is everywhere around."

Tea pickers were busy on every slope, gathering the fresh shoots. "They seemed a happy and contented race; the joke and merry laugh were going round, and some of them were singing as gaily as the birds in the old trees." The pickers were mostly women, with broad straw hats shielding their faces from the sun and large grass baskets slung across their backs—and perhaps even a child slung in front. The pickers stayed in the fields from early morning through dusk, from April through October, picking each bush every ten days. A female traveler, who followed Fortune's route in 1870, wrote:

> I am greatly struck by the number of girls whom we meet working as tea-coolies, and by the enormous burdens which they carry slung from a bamboo which rests on their shoulder. Each girl carries two bags thus slung, the weight of a bag being half a picul, which is upwards of 60 lb. Thus heavily burdened, a party of these bright, pleasant-looking young women march a dozen miles or more, chatting and singing as they go. . . . The tea-plantations are scattered over the hills, forming little dotted patches of regularly planted bushes. Here the girls and women are busy selecting the young green leaves, which they pick and collect in large basket-work trays of split bamboo.

Tea-picking women, the *caichanu*, were heralded in song and story as dainty, noble, and objects of romantic interest. China's long history of politician-poets ensured that throughout the various dynasties there would always be some formalized appreciation of the beauty of these women and the harsh conditions in which they worked. A popular analogy held that the tea-picking girl had all the purity and nobility of the tea she picked, and contained in such beauty was hardship. Tea's purity was personified by the virginal *caichanu*, and her diligence was embodied in the tea mountain's majesty, for tea picking was dreary, soul-destroying work.

The tea harvest did have some advantages for the women of rural China, however. It relieved them of the solitary sphere of work in the home and brought them into the world. If a *caichanu* on a hillside was subject to the harsh light of public scrutiny, where her morals could easily be brought into question, the tea harvest also permitted her a degree of freedom. Walking the hillsides with other women, she could escape for a while the tyranny of a mother-in-law and the confines of familiar walls. It was this seasonal liberty—so contrary to Confucian notions of familial right and morality—that invited both the notice of scholars and serenades. "The Ballad on Picking Tea in the Garden at Springtime" goes:

> Each picking is with toilsome labor, but yet I shun it not,
> My maiden curls are all askew, my pearly fingers all
>   benumbed;
> But I only wish our tea to be of a superfine kind,
> To have it equal their "dragon's pellet," and his "sparrow's
>   tongue."
> For a whole month, where can I catch a single leisure day?
> For at earliest dawn I go to pick, and not till dusk return;
> Then the deep midnight sees me still before the firing pan—
> Will not labor like this my pearly complexion deface?

Today tea plants are allowed to grow only to waist height, and each bush is "tabled," a process that leaves it looking as if its entire top half had been lopped off. This makes for a wide, low base for convenient picking, and bushes are kept in neat, efficient rows. But in Fortune's time they grew virtually wild over a hillside. Fortune was stunned by the sheer labor involved in tea picking. It was not the bending or the hot climate or even the high altitude that made the work so arduous but, rather, the sheer volume of tea picked. If the tea bush were a Christmas tree, pickers would take leaves only from the bough where the star is placed, the very tip, and perhaps from a few of the branches with ornaments on them. From each bush comes only a handful of leaves because only the two most tender leaves sprouting from the end of a branch release the gentle and mellow taste that becomes tea; the older leaves on the stem below taste harsh and sooty. A nimble tea picker can pluck up to thirty thousand tea shoots per day, which includes the time it takes to examine each shoot and make sure no stalk enters the mix. It takes about thirty-two hundred shoots to make a pound, so an expert tea picker might gather as much as ten pounds of green leaf a day. The ratio of picked leaf to dry is five to one, meaning five pounds of fresh tea leaves are picked for every processed pound for sale.

"The natives are perfectly aware that the practice of plucking the leaves is very prejudicial to the health of the tea-shrubs, and always take care to have the plants in a strong and vigorous condition before they commence gathering," Fortune noted.

Picking places terrible stress on the tea plant. A bush is picked every ten days, from late April, once the rains have stopped, all the way to October, when the rains begin. It is a repeated undermining of the growth process and a wound to the bush, but continuous picking also produces the best brew. Cultivated tea bushes develop a deep network of taproots to compensate for the constant pruning; the roots then push up a rich healing sap that fills the leaves with flavor. Pickers also clear the tea bushes of their fruit and flowers as they move across the mountain. Anything that distracts the efforts of the bush, such as forming fruit to distribute the seeds, detracts from the energy it needs to heal itself after picking to produce more shoots.

It was not a matter of simple academic interest to Fortune that he study tea of fine quality. The price of tea reflected how careful the picker was at the harvest, and a failure to pick discriminately would ultimately cost the farmer at market. If the East India Company planned to produce a premium product, it would need to follow these methods as well.

Along the rocky path, Fortune stopped an old peasant and asked for directions to the local temple where his retinue could stay the night.

The peasant laughed at the request. "There are nearly a thousand temples in the Wuyi Mountains," he replied.

Fortune walked on toward a large temple lying at the foot of the mountains. The outer walls were imposing, but inside were lotus ponds, arched bridges, and screened promenades. The arrangement of the temple buildings was perfect: Its wide courtyards were aligned to the compass and yet maintained perfect sight lines to the lakes and rivers below and to the treetops and mountains.

Fortune had entered a temple of Buddhism, a religion that exalts Nature and the life force within, and the excellence of all living things. The Buddha lived in India at the time of Confucius. He preached that all beings pass through a series of lives, a cycle of endless reincarnation, paying for the sins of one life with a good deed in the next. Attachment to any one world is a cause of suffering, he said, when we are only ever passing through. Buddha practiced a series of spiritual exercises to end his attachment to this world so that he could focus instead on the one to come. With discipline, he said, we can escape the prison of the self and the cycle of rebirth to enter Nirvana—nonbeing.

In the view of the Buddha, every tree and plant was an honor and a gift, to be treated with veneration and tended with care; in that belief, the monks cultivated the temple grounds lovingly. The trees were all pruned and carefully planted in groups. Beauty was ritualized; nature was choreographed. There were also, unsurprisingly, tea shrubs to be seen in every direction, for tea was the most contemplative of nature's gifts. Beyond the monastery walls, the forest was entirely untouched, with old trees reaching toward the sky. "In this respect these priests resemble the enlightened monks and abbots of the olden time, to whose taste and care we owe some of the richest and most beautiful sylvan scenery in Europe," Fortune noted.

A young boy of six or seven who had just received his robes as a novice monk was sitting under a temple porch when he caught sight of the tall mandarin approaching. Noting the stranger's appearance, and perhaps even noticing the strangeness of his bearing, the boy ran across the courtyard into one of the smaller buildings.

Fortune was weary from the hot morning's climb, and his silk robes were heavy with sweat. He wandered into the long reception hall, lined with carved chairs and latticed screens, to take a seat out of the sun and to await his official welcome.

Sing Hoo proudly entered the hall to negotiate with the high priest. Fortune's stay was easily negotiated: Of course he would be received; strangers always were. The mandarin was a man of honor and so would be given the finest room, tobacco, rice, and tea that the monks had to offer. The priest then dispatched the boy with orders to make their esteemed guest comfortable.

He returned with a small iron pot of tea; it was a ripe oolong, with the aroma of orchids and peach pits. The boy palmed the cup, only slightly bigger than a thimble, and bent low as he handed it to Fortune.

"And now I drank the fragrant herb, pure and unadulterated on its native hills. I had never been half so grateful before, or I had never been so much in need of it; for I was thirsty, and weary."

The monks prepared a banquet lunch to greet the noble stranger. According to Chinese custom, then as now, it is considered a great honor when a host overfeeds a guest. The monks' meal was generous and lavish, with the best of the mountain's summer harvest: lotus root, mushrooms, pickles, cabbages, and beans. They drank liberally, and although Fortune was not usually fond of Chinese liquor (which he judged to be "rank poison"), for once he found it "agreeable," much like "the lighter French wines."

The entire monastery was in attendance at the meal. One monk had a face ravaged by smallpox, which did little to enhance Fortune's appetite; another had a face beatified by meekness and prayer, which made Fortune glad to be in faraway Bohea. Although Fortune could just understand the monks' Chinese, he thought it wiser not to speak in the temple. Sing Hoo, Fortune noted waspishly, "was quite competent to speak for us both." And yet, despite the distance between his world and the monks', he felt warmly welcomed and generously treated. He felt at home: "We were the best of friends."

Sing Hoo, too, enjoyed a distinguished position among the humble priests, for he was a well-traveled man. He had seen the imperial wonders of China. He could describe the emperor's yellow robes, the pleasures of the Forbidden City in Peking, the marvel of the Grand Canal, and the Great Wall in all its glory. The monks were engaged, provoked by thoughts of the wider world despite their remote life in Bohea. Indeed, Fortune was struck by how little the supplicants engaged in prayer and how involved they were in daily tasks. It seemed to him that his hosts paid "more attention to cultivation of tea than to the rites of their peculiar faith."

Tea was indeed central to monastery life. It was served at all times of day and at each meal. Within the temple grounds every vista included tea bushes: Tea was in the hedgerow; tea was at the gate; tea was an ornamental decoration. Fortune had arrived at the height of tea-picking season when the second flush had just come on, and picking baskets woven of bamboo were scattered among the bushes where the pickers had dropped them in the fields before leaving to eat and rest. In every courtyard there was a wide, flat bamboo withering tray, full of the morning's plucking, drying in the sun. Tea was a religion to these monks, a holy charge, and tending tea was a form of meditation.

Wuyi's monks were also assiduous record keepers, much as monks in the monasteries of Burgundy had noted for centuries which parts of a hillside grew the healthiest vines and faithfully documented the yields of hundreds of harvests. Fortune, too, took profuse field notes on latitude, longitude, rainfall, and the consistency and color of the soil: rocky and well drained.

As much as Fortune believed that the fate of nations rested on his research, his work in Wuyi Shan would also affect how every pot of tea would be prepared in the future. From a factory worker and his morning brew to the housewife with her evening cup, each man and woman in England had an opinion about how to brew the perfect pot of tea, and as Fortune would become the de facto leading expert on the subject in the West, it would be his job to tell them definitively whether or not they were correct.

At first glance Fortune's attempt to apply scientific methodology to the task of preparing tea seems questionable, given that nothing more sophisticated than a "quiet palate" was generally considered the best way to appreciate tea's subtleties. Tea is not so much a thing as a cupful of effects, and as such does not lend itself to hard-and-fast rules and rigorous testing. Yet Fortune, ever the diligent shirtsleeves scientist, took notes and analyzed the simple steps behind preparing every cup:

> Boil water.
> Ready cup.
> Add dry leaf.
> Drink.

Boil Water

The first ingredient of tea—indeed, nearly the entirety—is water. Experts hold that the condition of water in tea making is of greatest importance: its precise temperature, how long it has been boiled before it is poured, and whether it is fresh from the source or stale. As much as Fortune paid attention to the taxonomy of tea, he noted the variations in the preparation of water.

Water could not be merely hot, he wrote, but must be at a boil. It must not be allowed to boil for too long, however, as that would release the concentration of air suspended in the water. Any tea made with overboiled water will taste flat, just as champagne without the bubbles makes for a white wine with little body. Yet for the delicate nose of green tea, the water must not reach a full rolling boil, either.

So how does one measure the ideal tea temperature? Fortune reported a Chinese rule: "Do not boil the water too hastily, as first it begins to sparkle like crab's eyes, then like fish's eyes, and lastly it boils up like pearls innumerable, spinning and waving about."

Ready Cup

In Europe as in China, there was a general preference for the pre-warmed cup. To this day the Chinese warm the cup itself with the first of the tea. An entire cup is poured and then discarded unceremoniously. The reason for this is that tea leaves are, like grapes or apples, produce in need of washing, which does not happen at any time during processing. In fact, the finest teas are often processed in a spectacularly dirty fashion. They are left to dry on the ground, in the dust, where they may be visited by rodents and insects, and then stored on a factory floor in open sacks. This first cup is said to be for the demons or "for your enemies." Pouring out the first brew also has the practical advantage of warming the cup. In Britain hot water was typically run through an empty teapot to heat the pot, which would otherwise cool the tea too hastily as it brewed.

It is said that this practice developed among the lower classes, who did not have servants to clean their teapots properly after each use. Despite the unflattering class assumption—that the lower orders were less than hygienic—prewarming the cup or teapot keeps tea lively longer. Cooling tea, according to the Chinese, is perfection. Cold tea, however, is a sin.

Add Dry Leaf

How much tea goes into the perfect cup? In Fortune's day there was so much adulteration in the tea exports to Britain, so much twig and stem padding the weight, that it was nearly impossible to predict how strong a pot would be. Then as now there was a general rule by which most tea was brewed: about one teaspoon per cup.

Oversteeping tea makes it "stewey," in the words of the trade. No self-respecting Chinaman in Wuyi Shan would dream of letting a hand-picked tea brew longer than it ought.

Fortune's investigations proved that good tea goes further, that less can be used to brew more. His conclusions also reflected a sensible economizing: If better quality tea costs half again as much as that of poorer quality but brews up twice as strong, it is preferable to purchase the finer tea and enjoy the experience more.

In his travels Fortune found much variety between the teas of different regions: the appearance of the leaves, the aroma, the color of the liquor, and the taste. Tea, like wine, has *terroir*, a flavor that reflects the characteristics of the soil in which it grows. Connoisseurs today enjoy these contrasts and seek them out. In some areas, in some pickings, the leaf grows large and flat because it faces the sun. Some areas grow only small-leafed tea. There is no perfect specimen, nor is there a grand unified recipe for brewing tea. A large tea leaf must be steeped for a long time, a small leaf for a shorter time: It is a function of the ratio between surface area and water. The smallest leaves, the so-called dust-grade tea, which are the component of most tea bags today, brew up the quickest.

Drink

Tea is a stimulant, albeit a mild one, a property that has rendered it the second most popular drink on earth, after water. Tea promotes mental alertness, happiness, and sharper perception. "Tea is of a cooling [*yin*] nature," Fortune reported. "And, if drunk too freely, will produce exhaustion and lassitude. . . . It is an exceedingly useful plant; cultivate it, and the benefit will be widely spread; drink it, and the animal spirits will be lively and clear. The chief rulers, dukes, and nobility esteem it; the lower people, the poor and beggarly, will not be destitute of it; all use it daily, and like it. . . . Drinking it tends to clear away all impurities, drives off drowsiness, removes or prevents headache, and it is universally in high esteem."

The distinctive taste of tea—mildly acidic, with a bit of salt, an astringent—is an amalgam of several different chemicals. Tea contains plant enzymes known as phenolics, which are produced when its leaves brown and are bruised. Phenolics have a brisk, lively taste, which contribute to the sensation of the brew's being stimulating but gentle. Theanine—a tea-based counterpart to caffeine—is an amino acid that straddles the line between sweet and savory. We know now what effects theanine and caffeine have on the body; the effects have made caffeine the most widely consumed behavior-modifying drug in the world. Caffeine is a chemical alkaloid, a base, which interferes with the way cells signal the body. It stimulates the nervous and cardiovascular systems. It raises mood levels and decreases feelings of fatigue, increasing attention and quickening reactions. It also affects the heart, raising the heart rate, dilating arteries, increasing blood flow, and raising the respiratory and metabolic rates for hours after it is consumed. Caffeine in great quantities results in nervousness, restlessness, and sleeplessness.

It is worth considering which drink, tea or coffee, is the most stimulating. The answer is black tea—but with certain caveats. Per pound, black tea has more caffeine than coffee—but where one pound of tea brews some two hundred cups, a pound of coffee yields barely forty. By the cup, black tea actually contains roughly half as much caffeine as coffee. Green tea, meanwhile, has one-third the caffeine of black, or one-sixth that of a cup of coffee. Medically, it takes about 200 milligrams of caffeine—or about two cups of coffee—to combat drowsiness and fatigue. That amounts to about four cups of black tea and twelve cups of green. Few of us have that amount of time or bladder capacity.

What the world has sought when it sips a cup of tea is a mild effect, a high with neither lift nor letdown, a calming alertness, a drink of moods. What Fortune found in Wuyi Shan was Britain's reigning temper: the thrill to conquer, but politely.

From the monastery Fortune was within a day's walk of the Big Red Robe bushes, the Da Hong Pao, source of the most rarefied tea in the world—and certainly the most expensive. The three 200-year-old bushes stood under the three characters *Da, Hong, Pao,* which had been carved into a rock face and were fiercely protected by the monks.

Legend held that the bushes had arrived in Wuyi after nine evil dragons ravaged the area, causing havoc, destroying crops, and ruining lives. Finally, an ancient and immortal god arrived to confront the menace and restore order to the countryside. A great battle ensued, the skies darkened, and the forces of good and evil were pitted against each other. The god destroyed the dragons one by one, and where each corpse landed, a tall peak formed. It is said that the nine karsts of Wuyi Shan are these dragons, frozen in a fighting stance. The river bent nine times around the nine dragonlike karsts, and the area became known locally as the Jiulongke, the Nine Dragons' Nest.

That victorious immortal, celebrating his success, wished to leave a memorial to the battle so the people of Wuyi Shan would not forget him or his good deeds on their behalf. High in the ridges of the tall black mountains, above the river and on a sheer cliff where they would be hard to pick, the god left three tea bushes clinging to the rock face. Being of immortal creation, the tea bushes seemed to emit a red light, as if perpetually catching the reflected rays of the setting sun.

In the time of the immortal's victory over the dragons, an ancient Buddhist abbot, Tie Hua, looked up from his daily meditations and saw the three bright bushes illuminated by the light of the heavens. He was old and frail and could not reach the tea himself, and some said that no human hands could touch the immortal's gift. But Tie Hua was a man of many resources; he reached into his robes and brought out a monkey, his favored pet. The monkey swiftly climbed the precipice, picking his way up the rocks to the overhang where the immortal's bushes sparkled in the breeze. The monkey's small hand plucked the tip of the topmost branches, two leaves and a bud. An animal could not harm the trees' spirit, which was the mixture of heaven and earth.

Tie Hua gathered the tea leaves and returned to his monastery, where a revered local scholar was suffering great agues and chills. The man was bloated, racked with stomach distemper. He could not move and consequently would be forced to miss the imperial examinations for the civil service in Peking, his one chance for greatness and at bringing honor and wealth to his family as well as fame to his village in Wuyi Shan. The scholar was heartsick at the thought that he might not get to the emperor's court. But Tie Hua brewed the tea leaves and served the drink to the citizen scholar, who was immediately healed. The next day, entirely recovered, the scholar continued his journey to the north to sit for the imperial tests of knowledge, poetry, and character. His mind was alive; his senses were keen; he felt more able than he had ever been. The young man took first place among every scholar in the land.

Upon meeting the emperor, the scholar discovered that the empress was suffering from a similar illness: fevers, fatigue, and nausea. It had continued for weeks, and neither physician nor priest could cure her. The scholar had brought with him a pouch of tea from the immortal's three red bushes, which he offered in tribute to the emperor. On drinking the tea, the empress was instantly and permanently cured. The emperor demanded that tea from the first flush in Wuyi Shan—that is, the first bloom of spring—be sent north to Peking each year to cure the court's illnesses. Grateful and magnanimous, the emperor also sent the scholar back home with a generous gift of a large red silk blanket to protect the roots of the immortal's bushes in the coming frost. The tea has been called Da Hong Pao, the Big Red Robe, ever since.

A thousand years later the last of these tea bushes in the remote mountains of Fujian Province are guarded by armed men. They are no less precious now than they were when Fortune was wandering the tea hills. Locals say that the young leaves still glow red through the summer, and their dangling fruit still looks like garnets in the sun. The Da Hong Pao bushes are watered directly by the heavens, by pure rain filtered through the hard rock of the mythical dragons' skeletons. Each season the ancient bushes will put out about one pound of tea shoots. The first and second flush of the Da Hong Pao, the most powerful and sweetest crops, sell on the private market as the most expensive tea per pound in the world. At several thousands of dollars per ounce, Da Hong Pao is many times more valuable than gold.

Fortunately, nature provided tea farmers with a way of making Da Hong Pao available to the masses. Tea is easily cloned; any branch, cut and replanted, layered under soil, and left to sprout will soon develop a network of roots. The process of cloning plants is called agamogenesis, reproduction without gametes— that is, asexual reproduction. Cloning is a more expensive method of propagation than planting seeds, but it will produce direct genetic replicas of the parent plant. The technique has been used for thousands of years in farming to preserve treasured cultivars and rare strains. Genetic clones of the Da Hong Pao bushes have been likewise planted throughout Wuyi Shan.

Fortune was tireless in his collecting, bringing home many hundreds of seedlings from the hills of Wuyi Shan, offspring of the Da Hong Pao. He collected thousands of branches and layered them in soil inside the cases to produce clones, and he also hired small children to help him collect tea seeds, finding that a little money "went a long way with the little urchins." From the monks he purchased saplings one or two seasons old to sit beside the cloning branches in the glazed terraria of the Wardian cases.

He was also an avid collector of tea myths: It was said that Da Hong Pao tea, so inaccessible in the crevices of the high mountains, was at its best when not picked by humans. Fortune heard another tale involving monkeys: At harvest time peasants threw rocks at the monkeys who careened through the branches of Bohea's cliffs. The monkeys retaliated and returned fire at their assailants with anything they could reach—and in Wuyi Shan every other object is a tea plant. The monkeys accordingly threw handfuls of leaves and shoots, while monks would stand at the base of the mountains with baskets spread wide, waiting to catch the tea. Monkey paws are, of course, the perfect size for the delicate task of plucking only the top two leaves and a bud. "Monkey-picked" tea was prized for its purity. So, too, Fortune noted, was tea picked by virgins' hands. After a few hearty, well-brewed cups of Wuyi Shan oolong, the monks nodded in agreement: "Virgin tea is best."

Fortune walked out with the monks in the mornings, tracking the rhythms of the gardeners and the priests as they made their tea perambulations. He took note of their small-scale processing and drying, and of the regional differences between black tea and green tea. He picked with them, shadowed them, and followed their rituals, his notebook in hand.

As Fortune prepared to leave the tea hills, the monk who had served as his host offered him a parting gift: several rare plants and flowers. The botanist did not record which varieties were given to him, but it seemed to him as if the monk knew exactly what his guest was looking for. He chose specimens that Fortune had not previously collected or even known about. Delighted and embarrassed, the wordless botanist was happy to accept the plants. Each seedling was in perfect condition for transplanting, an ideal memento of the Wuyi hills. Fortune commented that the monk's gift "increased my store [of new plants] very considerably," and he was deeply touched by the gesture.

However, his espionage could hardly be sustained for an extended period of time, no matter how skilled a spy he proved to be. The trouble—as was the case with so many of the disasters that befell him—was Fortune's servant, Sing Hoo. As Fortune grew in confidence during his fruitful stay, so, too, did Sing Hoo. As the mouthpiece and sole negotiator for his master, Hoo began to take some liberties. His explanations of Fortune's origins grew increasingly elaborate. Rather than helping him maintain a low profile, Sing Hoo raised it and then embellished it in order to bask in the reflected glory of his master's perceived stature. According to Hoo, Fortune was no longer just a mandarin from a distant province whose biographical details were rather mysterious. Now he had become a very great man, a rich man with many wives, a venerated warrior and respected leader, from Tartary in Central Asia. He was a descendant of Genghis Khan. Fortune was a man whom all should approach with humility and "anxiety," since he had the confidence of the emperor. Fortune's newly invented personal history became more than a little uncomfortable for him, but he accepted the increased admiration of his fellows "with the utmost politeness."

Upon hearing that the monastery's honored guest was a man of such esteemed status and wealth, an ancient monk came to Fortune's quarters. Bent under the weight of his robes, the lines in his face like the characters of some ancient script, the man looked as old as the limestone peaks around him. He moved slowly and randomly as he made his way through the monastery's corridors. Weak and simple, he was "apparently in his second childhood."

As the ancient monk hobbled through the doorway to Fortune's room, he kicked off his thin slippers and immediately began the kowtow, the ritual of nine forehead-to-floor bows—a sign of submission and respect in imperial China. Knees down, hands down, head down, ochre robes splayed on the floor like flower petals. Then up again, each move a symphony of ancient joints and creaking sinews. "I raised him gently from this humil-iating posture, and intimated that I did not wish to be so highly honoured," Fortune wrote.

It was Robert Fortune's only occasion for self-reproach in nearly two years of stealing secrets from China. He felt for a mo-ment a twinge of something that could only be called shame: "I nearly lost my gravity."

13

Pucheng, September 1849

After packing seeds and shrubs, Fortune was a wanderer once again, on his way from Wuyi Shan to the Fujian coast. He "bade adieu to the far-famed Woo-e-Shan, certainly the most wonderful collection of hills I had ever beheld." He had every reason to believe that he and others would one day return to claim more tea from its majestic homeland. "In a few years hence, when China shall have been really opened to foreigners, and when the naturalist can roam unmolested amongst these hills, with no fear of fines and imprisonments to haunt his imagination, he will experience a rich treat indeed."

The return journey to Shanghai and civilization would be easier than the trip out. Rather than retracing his route through high ground, Fortune headed directly east toward the Fujian coast, where the mountain passes were lower and the road more certain. A straight line would take him to the seaport of Fuzhou, where he had once fought off pirates. He was following the traditional tea route, the same path the leaves of Da Hong Pao took to the world market. It was by no means a simple trip, however, for on the way to the coast he would encounter the dangers endemic to the Chinese opium trade.

One night a loud and terrifying ruckus broke out; angry shouting and arguing pierced the stillness of Fortune's room at an inn. Among the voices he could just make out the rasping tones of his chair bearers and the more polished Chinese of his servant. The argument grew louder and more vehement, and Fortune imagined all kinds of terrors invading the night. "I feared they were seizing my servant with the intention of robbing us, and perhaps of taking our lives."

Fortune finally threw on his clothes and reached for his small pistol. "Human life is not much valued in some parts of the country . . . and for aught I knew I might be in a den of thieves and robbers."

Sing Hoo was a man of many stories. For miles on the trail he bent Fortune's ear with grisly tales of his travels, the more frightening and more ornate the better. He spun yarns about noblemen robbed and murdered in their sleep, of merchants mutilated and travelers beheaded. Fortune tried not to listen, but as the days wore on and the novelty of travel faded, he began to take a kind of reluctant delight in the intricacies of the horrors and sheer inventiveness of Sing Hoo's imagination. But on that particular night, woken so abruptly from sleep, all Sing Hoo's grim narratives came flooding back to haunt him. Fortune could not shake off one particular image of a man who disappeared and was found later stuffed into his own trunk. He imagined himself, headless, pretzled into a Wardian case.

Running down a ladder and entering the inn's central courtyard, Fortune discovered that the source of the disturbance was none other than Sing Hoo himself. There were some eight or ten men surrounding him, including the chair bearers, who were half again Sing Hoo's size and shouting as if all the demons of Hell had been unleashed. Sing Hoo stood his ground in the middle of the scrum, his back to the wall, defending himself "like a tiger at bay." He was determined to fight off the mob alone—and in his hand he held his only weapon, a smoking incense stick. Sing Hoo was thrusting and parrying, every now and again poking the hot scented ember close to the faces of his tormentors. "The most adventurous sometimes got a poke which sent them back cursing and swearing rather faster than they came."

Fortune found the scene hilarious—Sing Hoo among the jackals—even while recognizing that his own safety was in danger and that the scene was "quite sufficient to alarm a bolder man." But he had been in China long enough by now to see the comedy of one stout man holding his own against ten, of his lone servant defending himself with a joss stick.

Fortune strode into the middle of the throng. The men quickly surrounded him. Reaching into his pocket, he brandished the small pistol for everyone to see. With a superior weapon before them, the men were immediately chastened and stood down.

Fortune was gambling—the pistol did not work. The moist climate of summer had rusted the loading mechanism shut. The chamber was empty and would remain so for the rest of the trip. He had wagered, however, that everyone would recognize the gun as a symbol of his power, while no one would be able to discover its uselessness.

"My chair bearers and Coolie, who had always treated me with every respect, immediately fell back in the rear, grumbling at the same time." He listened to their complaints. It seemed that the crowd was demanding from Hoo some cash they had been promised and had never received. Fortune's servant was at it again, on the take, making a squeeze, raking a little more off the top than the percentage Fortune allowed. He, as always, was the one who would pay the price for Sing Hoo's sins.

"Had I been an uninterested spectator, I might have enjoyed a hearty laugh at the scene before me; but I was in the midst of a strange country and hostile people, and, being the weaker party, I felt really alarmed."

Drawing himself up into his haughtiest mandarin stance, Fortune bellowed at Sing Hoo. Whatever the trouble was, whoever had started the argument, it was Hoo's fault, and he should be ashamed. How dare he jeopardize Fortune's trip with petty larceny? In a climate of thieves, why bring down the wrath of the locals upon them? The aggrieved coolies were all hardworking, honest men, by Fortune's reckoning; they had dealt fairly with him and had carried him for miles on end through the torturous vertical terrain. Fortune had no qualms about taking their side in this matter and publicly said as much.

The money in question amounted to some three hundred Chinese cash, or about one English shilling, a sum for which Fortune could never have imagined risking his life. It felt as if he were breaking up a squabble between schoolchildren, but he was in genuine danger. So, with ten local witnesses watching, he flatly ordered Sing Hoo to pay the debt without delay. By publicly humiliating Sing Hoo, he placated the eight coolies and the onlookers for the time being.

Fortune was staying in a simple travelers' rest house, but he was in the company of thieves. Some "were evidently opium-smokers, from the sallow colour of their cheeks, probably gamblers, and altogether such characters as one would rather avoid than be on intimate terms with." Wherever there was opium, criminal interests could not be far away. Users lived on the fringes of society and often had a penchant for skulduggery and crime. Fortune's temporary home was, in fact, a so-called flower smoke den, or *huayan guan*, a seedy pleasure palace where men enjoyed poetry, women, and opium, in no particular order. Another common name for such inns was "husband and wife dens," but as a Chinese scholar noted, "In reality they seduce the sons of good families. They are a place for secret adultery." The inn was one of many houses of ill repute along the route to the coast, for since the end of the First Opium War and the resulting institutionalization of the opium trade, Shanghai had become the center of drug commerce in the Far East, and addicts from throughout the country made pilgrimages to the coast to assemble a fix.

Opium is derived from *Papaver somniferum*, the poppy of slumber, an annual that grows in the mountainous regions of Central Asia, much of which was at the time in the hands of the British Empire. The opium poppy has white or purple flowers, reaches a height of 3 or 4 feet, and has a solid cylindrical stem that bends and droops under the weight of a bud but stands erect when the flower is in full bloom. In the center of the poppy is a large globe-shaped seed capsule that is covered by a papery skin. When this seed pod is sliced open, it exudes a sticky sap that is collected, drained, dried, and then kneaded into small balls or cakes. The principal active ingredient in opium is morphine, an alkaloid that deadens pain, produces euphoria, induces sleep and apathy, reduces fever, and relaxes muscle spasms. For thousands of years, from the time of Homer, opium has been used as a recreational and medicinal drug. It can be smoked, drunk, eaten, injected, or rubbed on the skin. In addition to morphine, opium also contains codeine, another pain-killing alkaloid.

By the middle of the nineteenth century millions of Chinese— it was estimated to have been one in every three adults—were opium addicts. The affliction was so widespread that the once prosperous economy of the country effectively went bankrupt for nearly twenty years. By Fortune's day imports of opium were rising 20 percent a year; forty-eight thousand chests were imported from India in 1845, costing $34 million ($962 million today), and rose to sixty thousand chests in 1847 on revenues of $42 million ($1.1 billion today).

Then as now, the population of China was largely made up of peasants, and it was they who were most likely to become opium addicts. On its first introduction to China, in the Ming Dynasty, opium was seen as a court luxury that came from abroad. The Dutch began to bring opium to China out of Jakarta about the same time Europe was discovering tea, also via the Dutch. (Then, as now, the Dutch made terrific drug peddlers.) An early scholar of Chinese medicine wrote that opium "tastes bitter, produces excessive heat and is poisonous. It is mainly used to aid masculinity, strengthen sperm and regain vigour. It enhances the art of alchemists, sex and court ladies. Frequent use helps to cure the chronic diarrhoea that causes the loss of energy. . . . Its price equals that of gold." Once opium became popular at court, primarily as an aphrodisiac, it was taken up by the aristocrats, scholars, and officials of the middle class. By the mid-nineteenth century its pleasures had found their way to the lower classes, where coolies, chair bearers, and boatmen—the people Fortune encountered daily—all benumbed their hard lives with a drug that produced a sense of calm and ease. As opium use became increasingly widespread, China's officials denounced the taverns and brothels where it was used by the working masses of China.

The consequences of opium smoking, which was considered a luxurious indulgence among the upper classes, led to vast social ills in the lower orders. Wrote a traveler: "Smokers while asleep are like corpses, lean and haggard as demons. Opium smoking

throws whole families into ruin, dissipates every kind of property, and ruins man himself. . . . It wastes flesh and blood until
the skin hangs down in bags and their bones are naked as billets
of wood. When the smoker has pawned everything in his possession, he will pawn his wife and sell his daughters."

Opium made for a weak and lethargic workforce, a population
that consumed more resources than it produced. In particular it
made a mockery of the emperor's army, and a number of contemporary reports decried its use among the military: "Many Cantonese and Fujianese soldiers smoke opium; there are even more
among the officers. They are cowards and they have spoiled our
operation. They are really despicable." With an army in thrall to
opium, could there be any wonder that the Qing Dynasty would
soon lose its hold on power—or that the Taiping rebels would be
so victorious? Another scholar complained, "Although there are
more than 10,000 [soldiers], seven out of ten are Guangdong natives. They are cowardly and not used to marching in the mountains. Plus most coastal soldiers are opium smokers."

Drugs such as opium and tea were the first mass-produced,
mass-marketed global commodities; everything and everyone
these "stimulants" touched, from the producers to distributors to
consumers, were altered in their wake. The global drug trade, in
which England and China were deeply enmeshed, produced new
leaders, new governments, new companies, new farming practices, as well as new colonies, new modes of capital accumulation,
and new modes of transport and communication.

From an economic and imperial standpoint, opium was mi-
raculous. It found new markets and customers almost effortlessly
and took up little room on the merchant fleets colonizing the
world in the first wave of globalization. From its earliest days
opium was a mode of currency that made the Far East trade run

fast and trim: It was lightweight, easy to pack, and fetched a high price.

Opium, like other drugs such as tea and coffee, and like sugar, was good for the empire. Where the English had been trading for their breakfast tea with silver and racking up a crushing balance of payments problem, the growing opium trade quickly reversed the imbalance in England's favor. China's silver payments to Great Britain were $75 million between 1801 and 1826 (about $1.3 billion today), but the outflow increased to $134 million between 1827 and 1849 (about $2.9 billion), all as the result of the opium trade.

It was through drug-based commercial enterprises such as the tea and opium trades that Britain became the greatest of all hegemonic empires. The British campaign to sell opium in China was tremendously profitable. It brought in £750,000 in 1840 (about $3.8 billion today) and rose to £9.1 million by 1879 (about $22 billion). Britain's all-conquering naval fleet was able to be constantly improved with newly minted capital from the sugar, tea, and opium trades. Without opium the India trade would not have flourished, and without India Britain's post-Napoleonic global ascendancy could well have collapsed.

As Fortune's party settled down for the night, he could still hear the resentful clucking and cursing of his laborers. He returned to his bunk, but sleep was beyond him. Through the rotting floorboards, opium smoke wafted into the room, thick and heavy, clinging to the floor, swirling over his luggage, mixing with the smells of mold and damp, infusing the space with the sickly sweet scent of burnt sugar. Although overpowering, the smell was oddly seductive. As Graham Greene later said, it "was like the first sight of a beautiful woman with whom one realizes that a relationship is possible."

In the dining room below, a group of men, including Fortune's chair bearers, reclined on a large bedlike sofa, called a *kang*, with a lit lamp at the center of the room. One of the bearers leaned over the *kang* toward the fire to warm a ball of opium, which gurgled and boiled as the flame lapped at the sticky liquid. With a long spoon he stuffed the marble-sized ball of opium, now a hot, sticky mass, into his pipe and then inhaled in deep drafts. There was a loud pop, the sizzle of solids metastasizing into smoke. "The stewing and frying of the drug and the gurgling of juices in the stem would well nigh turn the stomach of a statue," wrote Mark Twain. The user leaned back again on his pillow, holding his breath, and then exhaled in a delirium of poppies and painlessness.

"What madmen might do under the circumstances—for madmen they were while under the influence of the drug—I could not possibly foresee," Fortune wrote. He stayed awake the rest of the night, playing out the possibilities in his head, enumerating all the ways he could die, much like the intricate tales of death and destruction that Hoo wove during their long days of travel. "This kept me awake for several hours."

After the altercation with the chair bearers, Sing Hoo slept with his clothes on, facing the door and waking at the slightest creak or disturbance in the night silence. Finally the opium smokers' voices died down in the room below. Seduced by the poppy, they had "gone off at last into the land of dreams."

At first light Fortune was up and his gear packed. He would push on and put all the high drama of the night behind him. He called for Sing Hoo to get a move on, to round up the others and get started. But the inn was entirely empty, with neither coolies nor innkeepers to help him with the load. Everyone had absconded under cover of night, never to do business again with the likes of Sing Hoo or his master, the strange mandarin. Fortune had mistaken opium smoking for outright mutiny. He was now in a faraway town without his retinue and with an angry and shamed servant his only companion.

There was nothing to do but seek help. Fortune ordered Sing Hoo to go to the nearest village to engage more men. He warned his servant not to take advantage of any new bearers, and Sing Hoo set out as directed, still stinging from Fortune's rebuke.

As the morning passed, Fortune was eager to put some distance between himself and yet another scene of insecurity and danger. He feared that the mob from the previous night was not only complaining to all and sundry about the terrible mandarin and his awful servant, but was now plotting retribution. But there was no sign of any new chair bearers or, for that matter, of Sing Hoo.

Finally, in the late afternoon, Sing Hoo returned to Fortune, alone and defeated. No one would work for him, not at any price. Sing Hoo's miserably low status sank even further.

But Fortune was having none of Sing Hoo's self-pity or obsequious apologizing. He announced that they would most certainly not be staying another night in the inn while the servant tried to rescue his reputation. They would leave immediately and travel in whatever daylight remained. Sing Hoo, who considered himself a high-ranking servant, would have to carry the entire load himself, like a miserable lower-class coolie. Fortune ordered Sing Hoo to strap the luggage together into one parcel using rope and bamboo. He would do the heavy lifting alone until they were far enough away for no one to have heard of the evening's fracas.

As the two men set out from the miserable opium inn, it

started raining and was soon pouring down in torrents. Still, Fortune insisted they walk on through the flooding streets. They became soaked instantly, feeling entirely sorry for themselves and completely furious with each other. They trudged on through the mud until the inn was behind them, a distant memory on another hill.

When they were a mile beyond the city walls, the bamboo with which Sing Hoo was carrying the load slung over his shoulders suddenly snapped in two. Everything Fortune owned—luggage, specimens, and every single tea plant—was plunged into ankle-deep mud. Baskets opened up, seeds spilled out, and all of Sing Hoo's grass cloth, which he had bargained so hard for, lay strewn in the filth.

Sing Hoo and Fortune were in the vast Chinese wilderness, surrounded by farmland, alone and very wet. No one could see or hear them; there was no one to bear witness to their frustration. Instead of being angry, Fortune pitied his sodden, exhausted servant. "I had not the heart to reproach him. . . . In the mud and water he looked perfectly miserable."

14

Shanghai, Autumn 1849

Back in the relative comfort of Shanghai, a guest in the cozy compound of Messrs. Dent, Beale & Co., Fortune sat down at a writing desk. He had received a package from the government in Calcutta. All the hard months of travel, his entire sense of success or failure in China, would hinge on the information contained in this communication.

Fortune eagerly broke open the seal on the envelope and pulled out page after page of official documents, seemingly an entire bookful of them. There was a mass of information: reports from company botanists and officials detailing the fate of his first shipment of green tea out of China. It was all neatly filed in reverse chronological order, clearly and painstakingly copied by hand.

The company had provided a summary of the information contained in these documents. It was news he was desperate to have, and yet the content beggared belief: Nearly all the tea plants had arrived dead in India. One year of Fortune's work, all the company's investment, had been completely wasted. It was as if he had not been in China at all that first year.

He put his head in his hands and tried hard to make sense of the news.

Then he began reading from the beginning of the file.

The plants had been sent out the previous winter, but by March there had been a delay, a detour and transshipment through Ceylon. Yet when his tea had reached Calcutta at the end of March, all still seemed to be going well. "Dr Falconer reports receipt of cases of plants containing 13,000 young plants. . . . The greater part of the plants were reported to have arrived in Calcutta in a healthy and thriving condition," a report stated.

Fortune continued to turn the pages of the documents charting the fate of his seedlings. They had gone upriver on the steamer to Allahabad, but they were delayed there, too, for the Ganges was low; they stayed there nearly a month. Had the cases been in good condition, delays would not have been such a problem, even after the month's wait in Ceylon, but it was clear from the letters that by Allahabad they were not in good order: "Many of the panes of glass in the cases, too, were broken," the missive reported. Fortune read on, his stomach turning. If someone had had the foresight to reseal the boxes, their contents might have lived. Or had a competent gardener replanted the seedlings into pots right then and there and tended the tea like houseplants on their way upriver, they might have made the trip successfully. Alas, his plants had likely been left on the loading dock of a company warehouse or factory, ignored until the rains brought the water level up and the boats could pass.

Fortune's jaw drew tighter. He could only shake his head as he made his way through the Revenue Department reports: From Allahabad the cases had been loaded onto an oxcart for the mountain gardens of Saharanpur, the company's experimental plantation in the Himalayas, where they arrived in mid-May.

"The first batch consisting of six boxes of plants reached Seharampore [*sic*] on the 14th of May in bad condition not more than 30 or 40 plants having leaves on them, but they have begun to show signs of improvement. The second batch of 5 boxes arrived in better condition on the 9th [June] having on the whole 41 plants in good condition in the 5 boxes."

After sending thousands upon thousands of young plants, Fortune could count only eighty healthy survivors in the mountains of India. It was a statistically meaningless number, a failure rate beyond all reckoning, the worst possible outcome.

The accounting of his first year looked very grim:

NUMBER CASE	NUMBER PLANTS
6	8 plants in good order
7	All dead
8	1 plant in good order
9	2 plants in good order
10	all dead
11	2 plants in good order and 2 sickly
12	6 plants in good order
13	8 plants in good order and 2 sickly
14	4 plants in good order and have thrown out some strong healthy branches
15	2 plants in good order
Box without number	5 plants in good order and 1 sickly

And the seeds? By early July some seven packets had arrived.

"The result has been an entire failure, not one seed having germinated. I lately removed a number of seeds from the beds to ascertain their condition and invariably found them to be rotten," Jameson wrote.

In the pages of correspondence that followed, Fortune read what seemed to be bureaucratic blame shifting. Calcutta officials lobbed accusations against the gardening men in the hinterlands.

The men in the provinces insisted that no one knew tea the way they knew tea—certainly not other botanists in Calcutta—and that their word was final on all subjects pertaining to tea. In the end, Fortune's first plants may have been doomed by collective corporate incompetence.

In his own notes he eschewed the self-aggrandizing revisionism that characterized official communication in British India. Fortune wrote drily and simply that he had not succeeded and that attempting to move plants to a new home and develop an entire industry out of transplants was a very difficult thing indeed.

> In the autumn of 1848 I sent large quantities of tea seeds to India. Some were packed in loose canvas bags, others were mixed with dry earth and put into boxes, and others again were put up in very small packages in order to be quickly forwarded by post; but none of these methods were [*sic*] attended with much success. Tea-seeds retain their vitality for a very short period if they are out of the ground. It is the same with oaks and chestnuts, and hence the great difficulty of introducing these valuable seeds into distant countries by seeds.

The time had come for further botanical experimentation.

Although Fortune's entire green tea haul had been rendered useless, it seems he had no fear that the company would recall him from China. Instead he remained sanguine, focused, and not the least bit apologetic. Nor did he indulge in any self-reproach whatsoever but instead concerned himself purely with solutions. He was confident enough in his basic gardening skills to know he could obtain better results.

He had in mind a new method of seed transportation, one that would force seeds to germinate inside the Wardian cases. He had concluded that the problem with the previous shipment was that he had exceptionalized the seeds, dividing the life cycle of the plant into two distinct moments, the living saplings and the inert seeds, and shipped them to India as entirely separate cargo. But the Wardian case was the safest place for all manner of plant life, no matter what its stage of development. Fortune remembered that when Ward had made his original discovery, he had looked into a sealed glass jar and seen seeds sprouting several years after the jar had been shut. Plants weren't frozen in time in a Wardian case; they lived and grew. Seeds should not be isolated from a living environment and shipped in hemp sacks like rice; they, too, would thrive in a terrarium.

Fortune immediately initiated a Wardian case experiment: He planned to stow many hundreds of black tea seeds packed into the soil of the cases he was sending to the Calcutta Botanic Garden. This experiment was not so different from the conditions in which Ward had made his first discovery, but it would be conducted on a mass scale.

Among the specimens he was sending to India were a number of mulberry plants that had been gathered in the district where China's best silks were spun. Fortune had traveled through the silk district on the Yangtze as he made his way in and out of tea country and believed that India, with its thriving cotton industry, might well benefit from experiments in silk production. He planted the mulberry bush in the "usual way," as he did any other shrub of economic and scientific note, with enough space, soil, and light to be comfortably sustained on the long journey out. He watered the transplant, left the mulberry bush in the sun, and a few days later, after the soil had absorbed the water and the plant had adjusted to its confined new home, he scattered handfuls of black tea seeds—each the size of a marble—over the surface of the soil. He then added another layer of soil, about half an inch deep, over the tea seeds. Fortune had his box makers fashion crossbars for this case so that the earth would stay in place no matter what turbulence a sea swell or oxcart travel might bring. "This method will apply to all short-lived seeds," he observed, "as well as to those of the tea plant. It is important that it should be generally known."

Fortune's first mulberry box, planted with seeds from the Da Hong Pao, was opened in Calcutta and hailed as a complete success. Not only did the seeds survive, but they also had fully germinated on the voyage out and arrived in a healthy condition.

Falconer, the senior botanist in Calcutta, was delighted. A scientist's inventiveness in China had made up for a company gardener's incompetence in Saharanpur. Nature had triumphed over human inaction and bungling. But not only that, Fortune had made a great leap forward in the global imperial project of plant transfer—which was essentially the transfer of technology. If living plants as well as fragile seeds could travel overseas, then entire industries could be transplanted—not one plant at a time but one entire profitable species at a time. Fortune expanded the global exportation of knowledge and technology in a 4-foot-by-6-foot glass box. For an imperial power such as Britain, with a planet full of subject colonies just waiting to be tilled and planted for profit, Fortune's discovery was nothing short of revolutionary.

"The young tea plants were sprouting around the mulberries

as thick as they could come up," Falconer wrote to the company and Fortune.

Fortune, buoyed by his success, made up another fourteen Wardian cases using his new method. Knowing that the principle at work was essentially sound, he grew decidedly less meticulous about the layering for his next experiment. With a bushel of seeds on hand, he made a mixture of one part earth to two parts seeds and tossed them all in together, like so many raisins in a pudding. Then he spread the soil at the bottom of the cases planted with rows and rows of tiny young tea plants, only a season or two old. He now had every faith in the incubatory ability of the Wardian case. He believed his black tea seeds would survive the trip to India, and so he sent many of them.

On a ship, protected by glass, the seed/soil mixture produced thousands upon thousands of germinating black tea seeds, all of which sprouted abundantly on their way to India. There were too many healthy plants for Falconer to count.

"The success attending their introduction in Ward cases has been so great, that I would recommend the attention of Government to be confined to procuring seeds and sowing them as recommended," wrote Jameson. There was no longer any need to hunt for living plants, the tender yearlings small enough to travel but hearty enough to survive transplant; seeds could do the trick. The new method was better, "proved by the admirable condition in which Mr Fortune's seedling cases reached us. The plants so reared reached the plantation in full vigour of growth and were but little injured on being transplanted into beds."

Fortune's new seed-shipping method increased the yield over shipping live seedlings tenfold—"for every young plant there will be, on the arrival of the cases at their destination in the [Himalayas], ten available seedlings." Every tea garden in the Himalayas from that season forward would bear the progeny of Fortune's tea plants, a foundation for the Indian tea industry for generations. He had radically changed the job of a plant hunter, which henceforward might more accurately be called a seed hunter.

Seed selection and breeding constitute a large part of the cultivation of finer-quality teas. The difference in quality between what Fortune provided for the Himalayas and what had been growing there already (the teas in the first shipments sent to the London tasters) was vast. While processed tea is subject to the vagaries of climate and rainfall, harvesting and shipping, the raw material is of key importance, and Fortune improved the Himalayan tea stock beyond all measure. His tea seedlings would breed and crossbreed with the stock already in the Himalayas, the inferior seeds out of Canton as well as the native Assam variety. Through the next several generations of selection and breeding, Fortune's stock—bred for the highest Chinese tastes over generations (called the China *jat*)—would mix with the best qualities of the native brews, the heartiness and maltiness of Assam tea (called the India *jat*). The new hybrids would produce unique flavors, flowery, mellow, rich, and supple, the finest teas in the world.

15

Shanghai, February 1851

O n the quayside in Shanghai unfolded a scene of pathos and
heartache as eight Chinese tea experts made their farewells
to the land they knew and loved and to all their extended family
members. Although the mandarin from beyond the Great Wall
promised there would be tea where they were going, each of the
skeptical departing men believed only one place could grow tea:
China, the center of the world.

Mothers pressed packages of food on their sons. The men
bowed their heads in respect to their fathers, who had seen life-
times of hardship and looked upon the loss of a son as simply one
more in an endless chain. Wives, when the men were lucky
enough to have them, wept openly. Children clung to the legs of
departing fathers. The young tea makers bent down to kiss babies
in arms whom they would not see for many years to come. Finally
the men pulled themselves away and walked up the gangway to
the tender that would bring them to H.M.S. *Island Queen*, a
wooden side-wheeled steamer bound for Hong Kong.

Fortune was oddly unmoved by the grief of his travel companions. He found the departure instead "an amusing scene,"
given that the tea makers were such unsophisticated "inland Chi-

namen," afraid of the new and the different and very much less
worldly than the Chinese in the ports who were fully conversant
with foreigners. However miserable these men were to be "taking
leave of their friends and their native country," Fortune could see
nothing in their situation to pity.

At the mouth of the Huangpu River, in the deepwater port,
the *Island Queen* lay at anchor waiting for the manufacturers, the
tea-making equipment, Wardian cases, and Fortune. The ship
would depart for Hong Kong the following morning.

Fortune walked up the gangway after his new workforce. He
was leaving the mainland for what might, as far as he knew, be
the last time. He had completed his last task for the East India
Company: finding and engaging Chinese experts who were
willing to follow him to India. He had amply stocked the tea
gardens of the Himalayas with new plants and seed stock. He
had sent cases full of plants home to the English sale rooms and
to Kew, and he had packed up a further consignment of porcelain, silk, trinkets, and other curiosities to sell at auction when
he arrived.

Fortune had also acquired all the necessary equipment that
would be needed in India to establish a tea trade: the ovens, woks,
and wide spatulas to fire the tea with as well as the farm equipment specially developed to cultivate it. For this task he had dispatched Wang and Sing Hoo to the various mountain districts to
hunt down "a large assortment of implements for the manufacture of tea." And finally he secured a collection of the perfumed
plants that Chinese manufacturers used for scenting teas when
they were packed, such as jasmine and bergamot. Samples of
these scenting agents were packed and shipped with him and the
tea makers to India, with the names labeled in Latin as well as in
Chinese, and a loose Chinese transliteration beside them. For-

tune felt satisfied that he had done his job well. "Everything had succeeded beyond my most sanguine expectations."

Fortune then said his good-byes. He made the rounds of the international community in Shanghai, collecting good wishes and returning borrowed items. There was no sadness in these farewells; the diplomats and traders of the Far East were accustomed to parting from friends. "Nothing therefore remained for me to do except . . . proceed on my voyage to India."

It had been a great deal more difficult dealing with people than with plants. Fortune had hired true experts from the deep tea country to instruct the Indian *malis* in the proper planting and processing of the new crop. Finding willing employees had not been easy in the atmosphere of danger and distrust following the outcome of the First Opium War. Men from the Chinese interior were particularly wary of foreigners, having heard chilling tales of their barbarism. Fortune made the task even harder for himself by refusing to hire anybody but the sons of tea growers, men who carried with them the knowledge of generations. "Had I wanted men from any of the towns on the coast, they might have been procured with the greatest of ease. . . . But I wanted men from districts far inland, who were well acquainted with the process of preparing the teas."

Absconding with plants was one thing; absconding with men another altogether. "The Chinese authorities have always watched with the most jealous eye any attempts made to export the tea plant; and any endeavour to procure good tea makers would assuredly be foiled or greatly delayed by specious difficulties," advised an official in Calcutta. Fortune, heeding the warning, did not recruit the men himself. If he were caught enticing Chinese natives away from their homeland, he would most certainly be put to death for kidnapping, which would probably spark an international incident.

Within a few months of setting out from Shanghai, Dent's compradors had come through with true experts for Fortune, masters of their art who would also be prepared to teach it. The compradors assembled six tea men from the same districts in which Fortune had been tea hunting over the previous three years. The men were young, obedient, and willing, and each had signed on for a three-year term in India. The men "looked up to me with the most perfect confidence as their director and friend. While I had always treated them kindly myself, I had taken measures to have them kindly treated by others," Fortune wrote.

The compradors had also secured Fortune two men who were expert in the making of tightly sealed lead boxes for the shipment of tea. Proper packing would help preserve quality and remedy what was becoming a rather dismal "deficiency of fragrance" problem in Indian tea. "By the London Brokers the black teas sent home have all been declared to belong, as stated, to the class of fancy teas, and to be more or less faulty in aroma," reminded one dispatch. "This could scarcely be otherwise expected."

Fortune himself had met the tea experts he would be escorting to India only days before their departure. He took no direct part in their recruitment, and his notes on how this was achieved remain vague. He simply dispatched agents, hired through Beale's compradors, to the Chinese countryside and let them do their work. These compradors—men who worked for the European trading firms as purchasers, negotiating with the emperor's trade envoys—were known quantities to Fortune; he had used them for the past three years of plant hunting and in his previous three years for the Royal Horticultural Society. He felt he could depend on them to find suitable experts and negotiate a fair price.

In imperial China's millennia-long history, the country had never officially recognized emigration. Every Chinese citizen was considered both the subject and the property of the emperor in Peking, and consequently going abroad was viewed as a theft from the Son of Heaven. For centuries China had forbidden its people access to the ocean, even to fish. It was among the most important duties of local officials to stop emigration. The ban on foreign travel was both a functional and a foundational part of Chinese culture. The Qing court feared invasion and so prohibited any contact with foreigners. Laws prevented political contacts with outsiders; there would be no free marketplace of ideas. The law in fact reflected traditional Confucian values, which held that to abandon parents, connections, and ancestors was a shameful act.

Despite the Chinese government's prohibition on emigration, a vast and thriving trade in sending Chinese workers abroad had developed during the late Qing empire. As the African slave trade wound down in the latter half of the nineteenth century, the coolie trade replaced it on the global exchange in cheap labor. When the traffic in African slaves officially ended for Britain in 1833, the empire was unable to find workers for its sugar colonies. The cost of abolition was high for Britain and showed up on the balance sheets of its merchant ships; simply stated, bodies were needed to fill a production gap. By midcentury, the discovery of gold in Australia and California had lured thousands of Chinese who could no longer eke out a living on the land as famines and floods devastated China. In the first few years of the American gold rush, some twenty-five thousand Chinese coolies would emigrate across the Pacific to California. By 1870, two million Chinese had found their way across the globe.

All too often, though, the only difference between an African slave and a Chinese coolie was that the Chinese possessed a contract.

"For Sale: A Chinese girl with two daughters, one of 12–13 years and the other of 5–6, useful for whatever you may desire. Also one mule," read a typical handbill at the time.

Coolies were enticed into coolie-ticket contracts by brokers using all manner of duplicity. Some were seduced by fantastic stories of the promised land to which they were heading, embellished with promises of free clothing, food, lodging, travel, and a fortune to be made. Some coolies signed on to repay their gambling debts. Others were sold by their families in the aftermath of clan wars, abducted as pirate bounty, or kidnapped in the middle of the night by crimps—thieves who dealt in human flesh. (The term "shanghaied" comes from the fact that many coolies were drugged, stolen, and shipped off to the flesh markets of Shanghai to be traded as chattel.) All coolie immigrants were put into holding pens called barracoons; they would wait there, locked up for months, until a full shipload was ready for dispatch to the New World.

The ship quarters for coolies were nearly as cramped as those on the African slave ships. A coolie's queue was cut off as a sign of his obedience to his new master, symbolically severing the ties of fealty to the emperor. His clothing was burned on arrival in Shanghai—he was charged for the cost of replacements—and he was then scrubbed with straw brooms to eradicate any lice and vermin he might have brought with him from his home. Once aboard ship the men stayed belowdecks and were seldom allowed—and were often incapable of coming—topside to breathe fresh air for the many months of travel required to reach the New World. The coolie ships were consequently hotbeds of dysentery, disease, and death.

Not surprisingly, under such conditions coolie mutinies were routine. In 1852, the *Robert Browne* sailed from Amoy for San Francisco with as many as 475 coolies on board. Once at sea and confined to the hold, the men were forced to sign contracts of labor; those who refused to cooperate were flogged. When the group's health began to deteriorate, the crew simply threw the sick and dying overboard. Nine days out of port, a mutiny erupted. The captain, two officers, four crew members, and ten Chinese died in the fracas. Days later, the ship put in at the Yaeyama Islands of Okinawa. The coolies were put ashore or escaped—there are several versions of the story, some of which were recounted under torture—and the *Robert Browne* returned to China empty, ready for another load of wretched human cargo. The coolies stranded in Okinawa later found their way to Canton, where they told their stories to missionaries and sympathetic diplomats.

Both the Americans and British in Canton as well as the Chinese investigated the circumstances of the mutiny for two years. Should the mutineers be hanged? Did the crew members deserve such terrible deaths? No resolution was ever reached. The Chinese were sympathetic to the mutineers; some suggested seeking the remaining crew who had "kidnapped" them and beheading them. Popular sentiment was stirred up against the coolie trade in China, and locals agitated for a revolt, which would send the foreigners a message that impressments such as this would no longer be tolerated.

The British were naturally incensed that their citizens, the captain and crew of the *Robert Browne*, had met such a horrible fate at the hands of the Chinese. Britain wanted justice for the crew, but the Chinese bureaucrats could not allow an official investigation that would pronounce judgment on the coolie trade, for to regulate the trade would be to recognize it. And so, as with prostitution, neither the Chinese nor the British could publicly acknowledge that such a trade existed at all. With this head-in-the-sand logic, the practice continued entirely unlegislated for another twenty-five years. Officially there was no such thing as a trade in coolies; therefore, it could not be a problem.

It was in the context of this brutal and tumultuous atmosphere that Dent & Beale's compradors went hunting for tea experts. These well-heeled middlemen of the coast spoke in pidgin, were loyal to the Westerners, and had been affiliated with the big merchant houses since their apprentice days. More important, the compradors maintained a rich network of contacts in the countryside. They could go where foreigners such as Fortune did not dare. They could purchase tea, porcelain, silk, and people—all at a good price. The compradors were the essential middlemen who made the China trade run smoothly, and they had been making a good living at it for centuries. Whether they were exporting coolies or tea experts made little difference to them.

Fortune was nervous about the way the business would be conducted, though. It would be improper for the company to be openly associated with anything as unsavory as the coolie trade, so all the contracts and dealings had to be conducted with the utmost formality and protocol. Using a comprador was a necessary risk; it kept the dealings one level removed from the British government, which continually worried about how such espionage would be received by the Chinese. "Being, too, a private individual, anything done by him to procure good men and tea seeds for exportation to India would not attract the attention of the Chinese authorities," wrote the government in Calcutta. Working on Fortune's behalf, the comprador was instructed not to cajole, caress, lie, trade for pirate booty, or buy an indentured tea expert. He could not conjure a story about a better life in a distant place to peasants already cast off their lands. The comprador had to be honest. He must also be secretive, lest factory owners learn of his plan and report him to local mandarins. The comprador would then be killed, and as a well-known employee of Dent & Beale's trading house, he would tarnish by association the name of one of the greatest firms in the Far East.

But Fortune was even more anxious about importing "experts" to India who might prove to be really no experts at all. The comprador's recruitment mission was thus doubly delicate. To entice him away from family and friends in China, the sums offered to a tea manufacturer had to be very great. Fortune's candidates were offered a salary upwards of 33 rupees, or about $15 a month ($415 per month in today's money) for a three-year term in India. Fortune's men were intended to be of a "better class" than the previously imported manufacturers and would be paid accordingly, "for it cannot be supposed that men thoroughly skilled in the cultivation and manufacture of an article to which so much importance is attached as tea and also of which the Chinese to a man believe themselves to be the exclusive possessors can be procured to live abroad for the small consideration of [less than] 33 rupees per mensem." The first two months' wages were paid to each man in advance as well as a stipend for "diet money" for the three months of travel to the Indian gardens.

The comprador convinced his prospective experts that the alarming rumors they had heard about "buyers and sellers of pigs" did not extend to the tea trade. He assured them that as company men they would be treated well, as professionals. An entire industry depended on their knowledge, and in sharing it they would gain a tremendous amount of face. The tea experts would have not only their autonomy but also power over others: over brown people, over white people, and over entire hillsides and mountains of tea. In the new land the experts could grow tea the way they thought it ought to be grown. They would be assigned to different plantations and encouraged to compete against one another, to produce higher yields and better-quality products for which they would even receive bonuses from the government, "to encourage the Chinese to exert . . . the full development of their skill and knowledge." Each season the manufacturer of the best tea of each description, green and black, would be publicly and monetarily rewarded.

Each man was presented with two copies of the standard contract, one in Chinese and one in English:

> I [Name Here], a Chinese Tea maker, hereby engage to proceed to [the Himalayan Gardens] to manufacture tea in the Government Tea plantations on a monthly salary of 15 dollars or Rs 33–2 commencing from [date] and I bind myself to serve for a period of 3 years. I further engage that I will [work] diligently as a tea cultivator or in any other manner in which I can be useful and failing any part of the engagement I shall be liable to pay a fine of one hundred dollars to my employers. I acknowledge having received from [Mr Fortune] on the part of Government an advance of two months' wages or 30 dollars, etc, etc.
>
> Witness signed in Chinese

The terms of this contract seemed to be generous enough for the time and the place, except for the provision of one of the clauses: that a man who earned $15 a month would pay a penalty of more than six months' salary should he fail to perform his duties for any reason, including illness. As munificent as the company might have felt the contract was, the tea makers' employment was unquestionably a form of indentured servitude. Whether or not the Chinese under contract to the company understood this aspect of their being hired is not documented, but the record does show that at least some of the early Chinese experts were displeased with the terms.

The Himalayan gardens had in fact already been employing a handful of Chinese tea experts, working for Jameson, the incompetent overseer of the government's tea plantations. Many of these men had come from the disbanded Assam Company; others had been imported directly from China. "I myself well remember the arrival of the Chinamen . . . in June 1843," wrote a government functionary in his diary. "Ten Chinese Tea-bakers amuse the *puharree* [turban-wearing] population by their strange figures and stranger propensities." He notes that among the tea manufacturers' peculiarities, at least in the eyes of the locals, was the Chinese love of pork. (The population of the western Himalayas was, by and large, Muslim, and pork was forbidden them.) However entertaining the locals found the Chinese, the company was greatly discouraged by their performance. Of the ten Chinese tea experts originally signed on, two had died by the time Fortune arrived in China. The rest were all from Canton, which produced a poor product by international standards. The tea makers were also, in Jameson's estimation, disagreeable and stupid. "And so ignorant are they as to be unable to make the common Black Tea exported to Europe," he reported in a dispatch.

When the company wanted to split up the Chinese group and disperse individuals to the many experimental gardens in the Himalayas, the manufacturers staged a coup, refusing to be separated. The Chinese tea makers also used this occasion to demand higher wages.

Jameson refused: "I communicated to them the orders of Government but all have refused to go unless their wages are increased and state that if Government insists on them going that they beg to tender their resignations." This was characteristic of his dealings with the Chinese: They seemed simply to have forgotten that there was no way for them to resign or alter their contracts in any way without also sacrificing six months' wages. The remaining Chinese manufacturers sought an increase of 7 rupees per month (roughly $90 in today's money). "And with this increase they offer to enter on another engagement for three years."

Rarely in company documents is there any record of the voices of the powerless or colonized people with whom the company had dealings. Yet the dispatches from the North-West provinces of India include a copy of this rather remarkable communication from the Chinese tea manufacturers to Jameson listing their demands:

1. We were ordered to remain at Almorah [a plantation] on a salary of Rs 33–2 annas per mensem, which we always have obeyed. To remove to any other stations we beg to say that if we get our pay increase of Rs 7 each in addition to that we now get we will go.

2. As we have been about 7 years in Government employment as no hope held out to us of an increase of pay, we object to go to Dehra and to Porree [other plantations in the western Himalayas]. We have no objection to go any where the Government like to

send us, but in the event of our leaving this station, we
request for more pay than we at present enjoy.

3. If on the other hand Government want to send us to
   any new station without giving us an increase we beg
   to resign our situation and hope that it will be considered and accepted.

4. If Government increase our pay Rs 7 per mensem we
   are working to enter into an agreement of three years
   to the effect that three of us will serve at [satellite
   plantations] Dehra, three at Porree, and four at Ha-
   walbaugh and to remain stationary for that period at
   the above mentioned places.

Patiently Jameson reiterated to the men that they were already
under contract and that this could not now be renegotiated; he
showed each of them the original agreement marked with his
very own chop, the stamp that stood for his signature. Although
they could see their names on paper next to the unintelligible
English writing, the tea makers flatly refused to be reassigned to
other gardens. Perhaps sensing that Jameson was not the most
forceful of authorities, they repeated that they would not be separated from their countrymen and sent alone into the wilds of
India. Their position was that they had been hired to work together at a single plantation and that employing them individu-
ally on different estates constituted a breach of the agreed-upon
terms. If the government could unilaterally amend a deal, so
could the manufacturers. They also insisted upon the validity of
their complaint that seven years of service was too long a period
to go without a raise, and the government ought to have been
sympathetic to their circumstances. Finally, after much debate in
the absence of a translator, Jameson arrived at a compromise and

agreed to amend the contract. He consented to most of the tea experts' new terms: The men would be assigned in pairs rather than singly, and at higher wages; in exchange they would stay for an extra three years beyond their original contract.

Jameson was ultimately rebuked for the increased expense to the enterprise. "The Lt Governor considers it necessary to urge upon you the importance of observing the strictest economy in all your proceedings. You must be aware that the success of the experiment will be in a considerable degree tested by its economical results," wrote the authorities in Calcutta, displeased with the idea that stealing secrets from a sovereign nation also meant overpaying the Chinese minions who implemented them. In fact, the tea manufacturers' wages were never so high as to be unduly burdensome to the company. The objection to their demands was more a matter of principle in that the Chinese were paid higher-than-market wages, which offended the British sense of fair play (even as the company remained unperturbed by the fact that the British in India were paid substantially better than the Chinese).

Ironically the original tea manufacturers had not been especially useful to Jameson's experimental gardens. By everyone's assessment they were lazy and poor workmen. They taught the natives and *malis* little about tea's proper cultivation, processing, packing, or scenting. As Jameson complained, "If the Chinese Tea Manufacturers are left to themselves . . . they would never be particularly interested in instructing natives in the art of making good tea (though they would willingly and readily show the process) nor would they, themselves, pay such attention to its manufacture and packing as is necessary to ensure its good quality." He cajoled, he insisted, he raged, but he could not get what he needed from the Chinese under his employ. They remained deaf to Jameson's request, however aggressive and impolitic, for teas of higher quality.

Yet they were essential to the British plan. Finally, Jameson conceded that a Western overseer was necessary to keep the Chinese experts and Indian gardeners up to the mark. An overseer would increase the cost of the operations, but Jameson could not be everywhere at all times. "To superintend the plucking of leaves, or rather to order the leaves to be pulled when ready, the presence of an European is very necessary [because] if they are allowed to remain too long on the bushes they become hard and are only fitted for making the coarser kind of tea and finally, in transplanting on a large scale the presence of an European overseer is of importance in order to see that it is properly done and also that the plants are regularly watered."

Of course, Jameson's constant watering of the plants contributed to the problem. Fortune's plants would have been better off left to the regular rains of the monsoons, as the Chinese experts had suggested. But Jameson was by now so frustrated with his Chinese experts that he would not listen to their recommendations—even when those recommendations were right.

Jameson, thoroughly frustrated by their obstinacy, finally lobbied hard for their replacement—an awkward argument to make, given how much they had already cost the government. "The Chinese manufacturers who were obtained some years since . . . are, in my opinion, far from being first-rate workmen; indeed, I doubt much if any of them learned their trade in China. They ought to be gradually got rid of and their places supplied by better men, for it is a great pity to teach the natives an inferior method of manipulation." Ultimately the government heartily agreed, noting that none of the original Chinese tea makers was "first rate."

In the chilly February air the emigrating tea experts stood on deck and faced the Chinese mainland, watching it slip farther and farther away.

"The boat was immediately pushed out into the stream. Now the emigrants on board, and their friends on shore, with clasped hands, bowed to each other many, many times and the good wishes for each other's health and happiness were not new, nor apparently insincere. Next morning the *Island Queen*, Captain McFarlane, got under way, and we bade adieu to the North of China."

By Fortune's count, 12,838 tea seeds were germinating en route to the Himalayas.

16

Himalayan Mountains, May 1851

Behind a bungalow in mountainous Darjeeling, rosy-faced children were playing on a hillside, their laughter echoing off the steep slopes. White magnolias flourished their cupped blooms toward the sky, and rhododendrons glowed scarlet on every rock face; there is nowhere on earth with more floral abundance to herald the spring. The children's father, a Scottish surgeon-major by the name of Archibald Campbell, bent down, trowel in hand, beside *malis* to dig a new home for some seeds and seedlings recently sent to the eastern Himalayas from the company's gardens at Saharanpur. In keeping with the ad hoc administrative style of the government in India, Campbell was the one man given the task of administering, planning, overseeing, building, policing, planting, and dreaming up the newest addition to the Indian empire: the mountain town of Darjeeling, which was home to twenty British families. If Robert Fortune can be said to be the father of Himalayan tea, then Archibald Campbell was its guardian.

The Darjeeling hill station was a real-life Shangri-La, charming and green, picturesque and pleasant. The Himalayas do not resemble Chinese mountains; there are no elegant karsts shrouded in wispy cloud but, instead, a line of enormous, rough-hewn, craggy peaks. Darjeeling is an island of beauty set in this landscape that is remarkable for the sheer diversity of its flora. Just below the rugged crests, streaked with snow and crowned with firs, are lush tropical valleys. Orchids grow on palms just under the snow line, while temperate hardwoods such as oak and birch mingle with jungle fruits such as bananas and figs. Drier than the western Himalayas, at nearly 7,000 feet above sea level, and shaded from the afternoon heat by mountains and mist, Darjeeling was a welcoming new home for Fortune's black tea.

Bordered by other small mountainous kingdoms—Tibet, Nepal, Bhutan, and Sikkim—the isolated mountain paradise had been only recently added to the growing mass of British India. Darjeeling's warring neighbors had passed the little district back and forth for generations until, in 1836, the rajah of Sikkim granted it to the British government as a token of friendship and fealty. Darjeeling was a concession or, more accurately, a bribe to the East India company: The rajah believed that if Britain ruled Darjeeling, it would not try to annex the greater kingdom of Sikkim in its quest to conquer China from the West. (Sikkim did in fact remain a semi-independent state until it joined India in 1975.) The company paid rent, or tribute, on the territory of Darjeeling for some years, until Campbell accidentally and providentially ended the arrangement.

Campbell was among the world's leading experts on all things Himalayan. In his service to the company he had learned the mountain people's language, published articles on their customs, and had carefully recorded all key statistics for the region: its geography, zoology, ethnography, and, of course, local agriculture. He was as avid a hunter and herbalist as he was a student of Ghurka and Lepcha rituals. Educated in Scotland, like so many in the Indian medical service, Campbell had been laid low several times by illness and so was put in charge of the company's government in Darjeeling. He was among the many enthusiasts who helped collect and catalog the opulence of the Himalayas for the professional botanists at Calcutta and Kew.

Quiet and unobtrusive is how people remembered Campbell. He was a perfect functionary, a good, solid company man, and he had the distinction of being among the first imperialists arrested for botanical theft. Although Fortune was in genuine danger in China, he survived his spying trips largely unmolested by those from whom he stole. Campbell, on the other hand, had only recently been released from a Himalayan prison for his botanical crimes.

While prospecting in the mountains for plants with the botanical eminence Joseph Hooker, son of the director of Kew and close friend to Charles Darwin, Campbell was arrested by the rajah of Sikkim in 1849. On an impulse he had followed Hooker to reach "the Snows" of the Tibetan plateau on a botanizing trip. Campbell had initially remained behind when Hooker departed, worried that prospecting in Sikkim would jeopardize his dearly desired promotion to resident to the Court of Nepal, but he ultimately decided that he could not forgo an expedition that was "the kind of opportunity for which you would give gold." He caught up with Hooker five months into his journey and surprised him one night in camp. His was the first "white face" Hooker had seen in months, and they reunited with "unspeakable joy." Campbell was a mine of information for the young naturalist; he shared a wealth of local botanical and geographical knowledge and provided Hooker with coolies, guides, and supplies. Mrs. Campbell had sent along plum puddings, mince pies, and English sherry to remind the men of home's comforts. But at the Chola Pass, in the eastern Himalayas, they were captured. Campbell was "seized, bound, treated with brutal violence, and called upon, at the risk of his life, to put his signature to whatever might be dictated to him." They heard rumors that their execution was under discussion at the Durbar, the Sikkimese court. Even their Lepcha porters were harassed and "barbarously treated." When news of their arrest reached Calcutta, European "outrage" was tempered by government anxiety. The company, stretched to the breaking point with border wars in the Punjab and Afghanistan, warned against taking "extreme measures" to secure their release. Troops were dispatched to Darjeeling in a display of company force, although the government had no intention of deploying them.

Finally, after six weeks of confinement, Campbell and Hooker were released. Sikkim was rebuked: The rajah's allowance for the annexation of Darjeeling was canceled, and he forfeited a section of Sikkim's southern territory. Hooker thought the punishment hardly fit the seriousness of the crime; he wanted a further show of strength from the company in the form of total annexation of Sikkim, but Campbell was happy to be restored to his family.

The two men remained friends for the rest of their lives. Campbell used the botanical trip to the high Himalayan plateau to his advantage, saying he "had gained that knowledge of its resources which the British Government should all along have possessed as the protector of the Rajah and his territories." In a characteristic display of administrative perversity, Campbell was made British Resident of the territory in which he had all too recently been imprisoned.

Comfortable in a carriage in which he was protected from the dust of the high Gangetic Plain, Fortune traveled to the government plantations in the western Himalayas. Having set sail in February, he had arrived in Calcutta on the fifteenth of March, spending a few weeks with Falconer and his Ward's cases, marveling at the state of his tea seeds. "The whole mass of seeds, from the bottom to the top, was swelling, and germination had just commenced." In late April, Fortune met Jameson at the Saharanpur gardens and was able to inspect the small tea settlements scattered around the hills and valleys. Most of these gardens were not yet two years old, and all had been planted with Fortune's seeds and seedlings.

The encounter was frosty, for although the superintendent was overjoyed at the eventual success of Fortune's tea, the tea hunter himself surveyed Jameson's gardens and found them wanting. The plants looked mangy and badly tended, and were showing poorly per acre.

The two men sat down in Jameson's bungalow, where Fortune was forthright about his views of the enterprise, indicating what he would write in his report to the company and to the government in Calcutta. He praised the scheme of leasing the tea lands to the local *zamindars* (which translates roughly as "caretakers" but refers instead to a kind of sharecropper). As rich as the mountain soil was, it would otherwise have been left fallow had Jameson not been so dedicated to promoting the *zamindar* scheme. It pleased Fortune to see that the Indian natives took so readily to tea growing. "I am happy to add that amongst these hills there are no foolish prejudices in the minds of the natives against the cultivation of tea," he told Jameson and the company. He also related a story about a *zamindar* approaching him directly to beg two thousand tea plants from him so that the man could begin growing tea immediately. "It is of great importance that the authorities of a district and persons of influence should show an interest in a subject of this kind. At present the natives do not know its value but they are as docile as children and will enter willingly upon tea cultivation providing the 'sahib' shows that he is interested in it. In a few years the profits received will be a sufficient inducement."

His was not a popularly held opinion, however, for many of the company men in the Himalayas balked at the *zamindar* arrangements and at any attempts to educate the natives so they would become practiced horticulturalists. "I regret that it should have been deemed necessary to make stupid pedants of Hindu malees [*malis*] by providing them with a classical nomenclature for plants. Hindostanee names would have answered the purpose just as well. . . . If these names are unpronounceable even by Europeans, what would the poor Hindu malee make of them? The pedantry of some of our scientific Botanists is something marvellous. One would think that a love of flowers must produce or imply a taste for simplicity and nature in all things," wrote a contemporary with scorn.

But beyond his approval of the *zamindar* scheme, Fortune had some severe words for the gardening decisions of Superintendent Jameson. In particular Fortune took him to task for irrigating tea. Fortune had never once, in three years' traveling in China, seen tea under flood irrigation, and his tea manufacturers said the method was used only for rice. Although tea cultivation was practiced with slight variations from region to region in China, no farmer there would dream of using such a procedure. The heavy rains that come with India's monsoons and the runoff from the annual glacial melt are sufficient to water mountain tea. In every experimental garden Fortune visited where there was tea growing in a valley, it was rotting. At least a third, and as much as half, of his early crops were lost to failed experiments in irrigation.

Fortune's report to the company on this misjudgment was emphatic:

> I have already observed that good tea-land is naturally moist, although not stagnant; and we must bear in mind that the tea shrub is NOT A WATER PLANT, but is found in a wild state on the sides of hills. In confirmation of these views, it is only necessary to observe further that all the BEST HIMALAYAN PLANTATIONS ARE THOSE TO WHICH IRRIGATION HAS BEEN MOST SPARINGLY APPLIED.
>
> Indeed I have no hesitation in saying that in nine cases out of ten, the effects of irrigation are most injurious. When tea will not grow without irrigation, it is a sure sign that the land employed is not suitable for such a crop.

Fortune's last shipment of Wardian cases had arrived in Saharanpur in stupendous shape. He counted no fewer than 12,838 living plants when he opened his cases, and many more seeds were germinating—too many to count. To Fortune, the plants seemed as "green and vigorous" as if they had never left their home ground. It was crucial that his last shipment, the crowning achievement of his time in China, not be lost to bad gardening and Jameson's faulty decision making.

Tea proved to be so suited to the Himalayas, however, that even the eighty surviving plants from Fortune's earliest shipment were now thriving. As Jameson and Fortune sat down to discuss the evaluation, those precious plants imported in the first season were about 4 feet high and "profusely covered with blossoms." They had taken well to their new ground and were yielding a good return of seeds, if not a high volume of tea shoots. They had produced several thousand offspring, Jameson reported.

While Fortune felt deservedly proud of his accomplishments, he tacitly acknowledged that Jameson had played at least some role in the tea's success. "The flourishing condition of many of the plantations is, after all, the best proof and puts the matter beyond all doubt," wrote Fortune to his superiors.

The Chinese manufacturers, meanwhile, were given "nice cottages and gardens." They were told they would be separated, divided up among the plantations, and they eventually were— although this caused consternation among the new recruits as it had to their predecessors. The Chinese all wanted a fellow countryman to talk to, but the government required one expert per plantation. Still, Fortune felt that the company was serving the tea manufacturers as best it could since "everything was done which could add to their comfort in a strange land."

On the morning of Fortune's last day in the Himalayas, the tea experts rose early and bathed. They dressed in their finest linen robes, celebration clothes saved for special occasions such as New Year's Day and the Full Moon Festival. In the dawn hours the appointed leader of the group, slightly older than the rest, a man of about Fortune's own age, stepped forward and presented him with a small token of their appreciation. Fortune makes no mention of what the gift was, but he felt he could not accept it. He thanked them profusely, extolling their virtues and generosity. "I told them how much I was pleased with the motives by

which they were actuated." A more sensitive man would have realized that to refuse such a gift would humiliate everyone, with a
corresponding tremendous loss of face.

He was still eager to help the tea makers in any way he could,
though, and readily agreed to accept a packet of letters from the
men—notes to wives, parents, and children. He promised to take
the mail with him to Calcutta and direct it to a steamer on its
way to China. He would do anything for the men, for "never,
from the time of their engagement until I left them in their new
mountain home, had they given me the slightest cause for
anger."

Fortune left the Chinese manufacturers in India with a heavy
heart. "I confess I felt sorry to leave them." It was as though he
was saying good-bye to all traces of China, leaving his last three
years of endeavor behind in India.

There is no hard evidence in East India Company documents of
the exact date of the earliest appearance of tea in Darjeeling. By
government accounts it would appear that the first tea seeds arrived in the Darjeeling mountains sometime in the early 1850s or
maybe as early as 1849—in other words, during Fortune's initial
travels on behalf of the company. But because his original shipment was largely a failure, it is unlikely that any of those seeds
found their way to Darjeeling. The founding tea stock in Darjeeling was green tea and likely came from Canton plants—all that
was available to the experimental Saharanpur gardens in the
1840s. The first black tea in Darjeeling, however, could not have
been from anything other than that sent by Fortune in those
Wardian cases layered with soil and seeds.

Plant exchange among Calcutta, Saharanpur, and Darjeeling

was commonplace. Campbell's mentor, Brian Houghton Hodgson, former resident to the court of Nepal, had stayed on in the subcontinent and retired to the hill country of Darjeeling. Hodgson, too, was an avid collector and amateur naturalist who kept an eye on the development of national industries throughout the Himalayas and maintained a lively correspondence with company botanists. Within the first year of Fortune's seeds and seedlings arriving successfully in India, it is very likely that the two lay naturalists cultivating Darjeeling, Hodgson and Campbell, campaigned hard to have the new seed haul subdivided and sent to the territory. If Fortune's seeds were not represented in the very first season's planting, they were most surely sown there by the completion of his travels.

Campbell had Fortune's tea neatly planted in rows on the face of a hill, a large tract of land that could be expanded as needed. There were many such experimental crops under cultivation in Darjeeling. Campbell was also concerned with cotton and even opium and other products that might be economically viable in the region. When Fortune's teas began arriving in the North-West Frontier over the course of the next two years, some portion was always set aside and sent onward to Campbell, along with other potential nursery stocks.

"Dr Campbell raised British Sikkim in ten years from its pristine condition of an impenetrable jungle, tenanted by half savage and mutually hostile races, never previously brought into contact with Europeans, to that of a flourishing European Sanatria [*sic*] and Hill Settlement, an international tribal mart of the first importance and a rich agricultural province," read one homage.

Today Darjeeling is considered the champagne of black teas. It has the finest brew, the most delicate floral nose, the richest liquor, and the most opulent amber color. At auction, Darjeeling teas fetch some of the highest prices in the world. It is almost impossible to buy a lot from the first flush of the Darjeeling estates; they are snapped up as soon as they come to market. As the Chinese say, tea is the essence of mountains, and the mountains that produce Darjeeling tea are the most magnificent on earth.

Within a generation India's nascent Himalayan tea industry would outstrip China's in quality, volume, and price. India, meanwhile, was to become an ever more critical asset to Britain—if it could manage to hold on to it.

17

Royal Small Arms Factory, Enfield Lock, 1852

Behind the gates of the Royal Small Arms Factory, Enfield Lock (RSAF Enfield), north of London, engineers and gunsmiths were experimenting with a new weapon to be used in India. Just as science and technology could revolutionize agriculture by making possible the movement of plants around the world, advances in weaponry would change the nature of soldiering—and perversely dissolve the East India Company's hold on India.

The company's fortunes had long depended on keeping tight control over the people and resources of India. It possessed an overwhelming superiority in arms and a monopoly on violence over Indians. The East India Company was potentate over the subcontinent, backed by the force of the Indian army. Ruled by Englishmen, the army was manned by Indians, with twenty-six thousand Europeans commanding two hundred thousand sepoys. From *sipahi*, Hindi for soldier, sepoys were Muslim, Hindu, Sikh, and Christian, and mostly recruited from the upper echelons of Indian society. The army oversaw a territory as large as the United States, containing a population of about 285 million. Without a private militia on the ground, the company could never have dominated India as it did or envisage using it as Britain's larder.

The experiments at Enfield gave birth to a rifle that came to be known as the Pattern 1853 or P53 Enfield Rifle. Based on this pattern, rifles would be manufactured according to a single standardized design that was soon used throughout the British Empire. In particular, the P53 would be sent out to India for use by Indian army regiments. Although gunpowder had been invented in China about AD 850, it had long been in general use around the globe to propel rockets, mortars, and bullets. But the technology behind weapons had not advanced substantially for hundreds of years. The gun used by the British in the defeat of Napoleon at the turn of the nineteenth century—the Brown Bess, a "smoothbore" flintlock musket—was still the most common one in the Indian army in the 1840s and early 1850s.

From the seventeenth century to the early nineteenth century, shooting a gun involved connecting a flame or spark with gunpowder in a confined space, creating an explosive pressure that would send a spherical projectile in the only direction it could go—out of the muzzle of a pipe barrel. The main limitation of these weapons was the amount of time it took to load one. Each reloading was a multistep process that required a soldier to pour a measure of powder, insert a lead ball and some wadding in through the muzzle end of the gun, and then with a ramrod tamp it all down to the end near the trigger. Ignition was provided by a flint striking steel behind the gunpowder charge, producing a spark near a hole leading to the powder chamber. A sprinkling of primer powder conveyed this fire to the internal gunpowder. Preparing a shot in this way took a seasoned and well-drilled soldier a minimum of fifteen to twenty seconds per shot.

Even in its day the Brown Bess was regarded as relatively inaccurate and unreliable. The problems with accuracy stemmed from the generous fit between the barrel and the ammunition. Since Bess tended to become fouled by gunpowder residue after a use or two, making it harder and harder to load, the barrel was designed to be somewhat larger than the projectile inside, which was a simple round lead ball. But this also meant that a musket ball had a lot of space around it inside the barrel, which caused it to bounce and bob through the barrel on its way to the exit. Gases from the explosion would also escape around the ball, slowing down the force of its propulsion. Being a smoothbore gun, Bess lacked rifling in the barrel, which also affected its accuracy. (Rifling—a groove cut in a spiral inside the barrel—helps impart a spin to the projectile so that it flies straighter and flatter.) The flint-on-steel ignition mechanism was also of some concern because it worked under fair conditions but tended to fail under damp conditions—a major liability in India, where the monsoon season lasts from June to November.

Despite such shortcomings Bess worked well enough for the military techniques of the day. Indian soldiers were trained to stand in lines perhaps two or three deep and to fire four rounds a minute at targets 50 to 65 yards away. When a phalanx of soldiers standing in a row fires volley after volley in the same general direction, whether or not their weapons are accurate or are certain to fire is less important than the speed of reloading; someone is bound to hit something in the field of fire. And when an opponent was in retreat or ammunition was low, the Brown Bess's 17-inch bayonet provided a more certain means of attack during a direct charge. For 150 years, during the entire period of the company's glory years on the Continent, the Indian army used this one weapon, backed up with cannons.

The longevity of the Brown Bess is not surprising: Armies have good reason to be conservative about replacing a known, tested technology with something experimental. Because armies are most efficient when equipment is standardized, with as little variation as possible, there is no simple way to phase in new and improved weaponry. But advances in personal weapons such as pistols and hunting guns were about to transform the standard soldier's firearm, making it more accurate, consistent, and deadly than it had ever been before.

The trick to improving the Brown Bess was to make a bullet that could grab on to the grooves of a rifled barrel and at the same time be quick and easy to load. An inventor at Enfield named Richard Pritchett thought a French bullet might offer help in mastering Bess's shortcomings and achieve these dual aims.

The French minié ball—roughly cylindrical but rounded on the leading end and hollowed out at its base—was more like the bullets of today. When it was discharged, the pressure of the expanding gases in its base would force the base to expand to fill the width of the barrel and thus grip the rifling grooves.

The early tests of the Pritchett experiment were greeted with enthusiasm at the Enfield Arms Factory. Whereas Bess could hit a target at 60 yards if handled by an experienced soldier, the new rifled weapons were accurate up to 600 yards. But the Pritchett bullet still required the marksman to go through numerous steps to load the gun, because the bullet did not come with its own casing, gunpowder, or priming charge. Moreover, because of the tight fit between bullet and barrel, the barrel needed to be kept greased so the bullet could be shoved smoothly into the chamber and pushed up tightly against the powder. To simplify loading, the bullet was packed in a paper "cartridge" that also included the necessary gunpowder and was greased on the exterior to ease its way into the barrel. Soldiers followed a precise, standardized routine for loading their weapons using these cartridges.

The P53 Enfield Rifle may have been a weapon superior to the Brown Bess in almost all regards and may even have been adaptable to India, had the company been more sensitive to the customs of the population over which it ruled.

The East India Company's Platoon Exercise Manual stipulated the correct method to load the P53, which was no different from the way Indian sepoys loaded the Brown Bess. Upon hearing the first command, "Prepare to load," a soldier placed his rifle butt on the ground, 6 inches in front of him. At the command "Load," the drill instructed: "First—bring the cartridge to the mouth, holding it between the forefinger and the thumb with the ball in the hand, and bite off the top; elbow close to the body. . . ." This standard procedure of "biting the bullet" was so fundamental a part of the routine that the phrase survives long after the procedure itself has been forgotten. With the cartridge and bullet still in his mouth, the soldier tore open the paper to gain access to the powder. He poured it into the muzzle and then shoved the bullet and greased cartridge paper (as wadding) after it, pushing it down to come in contact with the powder.

Grease on the paper cartridge was needed to keep the barrel slippery over repeated firings. It also protected the gunpowder in the cartridge from the wet, variable weather of India, since ammunition might be kept in storage for up to three years. Using a good, dependable grease became an integral part of the operation of the new P53.

The grease of choice for the company, because it was both cheap to manufacture and widely available, was a mixture of beef tallow and pork fat.

Had the Englishmen in charge sought to be insensitive to their Indian sepoys, they could hardly have been more successful than in their selection of a lubricant. Pigs are *haram*, outlawed, to the Muslim soldiers of northern India, while no high-caste Hindu will touch a dead cow, let alone bring one to his mouth. The animal fats in the P53's cartridges would therefore defile and debase every religious Indian who was commanded to use them.

The Enfield rifle was adopted and introduced to the Indian troops as a general issue weapon in 1856. The East India Company trained its infantry and musketry divisions first, and regiments and detachments were sent to the various arsenals and depots to receive instruction in the new weapons.

As soldiers were exposed to the new cartridges, rumors began to spread. It was said that the Enfield rifle was part of a mass plan on the part of the company to convert Indian troops to Christianity by rendering them impure, and thus forcing them to give up their caste status. After centuries of British domination in India, this seemed all too likely a plot to the angry Indian troops.

As the story goes, one January day a lowborn *khalasi*, or laborer, at the Dum-Dum arsenal near Calcutta said to a high-caste Brahmin sepoy who refused to sip water out of the same water pot, owing to his loftier ritual status, "The *Saheb-logue* [Europeans] will make you bite cartridges soaked in cow and pork fat, and then where will your caste be?"

Throughout the winter and spring of 1857 news of the tainted cartridges spread. On May 9 sepoys of the Third Bengal Light Cavalry flatly refused the order to bring the cartridges to their mouths and load their weapons. The rebels were court-martialed on the spot and sentenced to the unusually severe punishment of ten years' imprisonment under hard labor. For two hours the rest of the regiment stood watch in the afternoon sun as officers stripped the mutineers of their uniforms and paraded them naked, shackled in leg irons, through the army town of Meerut. Even the local prostitutes were disgusted and afterward refused their favors to anyone in the regiment. "We have no kisses for cowards," the women said.

That night several other sepoy regiments, from both the infantry and the cavalry, broke ranks and turned on their officers. The sepoys liberated the eighty-five rebels from jail, hailing them as heroes to their race. Then they burned the company's bungalows and offices. Every European was massacred on sight.

The cavalry retreated to Delhi, and for the next six months India would catch fire.

The P53 Enfield Rifle ignited a holocaust of murder, siege, brutality, and repression; women and children, Indian and British, were butchered, cities were sacked, and civilians were murdered by soldiers. The British refer to the summer of 1857 as the Indian Mutiny; Pakistanis and Indians refer to it as the First War for Independence. Whatever name is used, it was bloody beyond all precedent and it threatened the very existence of the East India Company.

The mutiny continued throughout the heat of spring. In the town of Kanpur (known as Cawnpore before 1947) in northern India, a former ruler named Nana Sahib brought the mutiny to its gruesome pinnacle. The company had recently deposed Nana Sahib according to the "doctrine of lapse," a policy by which a native Indian ruler was induced to "sacrifice" his territory to the company if he could not provide a legitimate heir.

When three hundred British troops, along with their wives and children, were imprisoned in the barracks of Cawnpore with no food and little water, Nana Sahib stepped in and offered the British safe passage downriver to a city still in company hands.

Starved, dehydrated, and diseased, some two hundred of them already dead, the British accepted Nana Sahib's offer. But as they boarded the boats for safety, the native crew members set fire to the boats' thatched canopies and suddenly jumped overboard. At the same time a volley of rifle fire erupted from the banks. Only four British men survived the escape downriver by swimming to safety.

The Indians gathered together those who had avoided the boat massacre and herded them into an open courtyard. They shot all the men. Some women and children survived for a time; some were carried away and raped.

The British sent a relief column, but it arrived a day too late; the sepoys had already butchered the remaining women and children in the courtyard. The "floor of the yard and the verandah and some of the rooms were bespattered with blood and the bloodmarks of children's hands and feet, women's dresses, hats, Bibles, marriage certificates, etc., lay scattered about," wrote one eyewitness.

Nana Sahib, the Butcher of Cawnpore, was said "to read Balzac, play Chopin on the piano and, lolling on a divan and fanned by gorgeous Kashmiri girls, to have a roasted English child brought in occasionally on a pike for him to examine with his pince-nez," a newspaper would later report.

When the British arrived, they ordered every mutineer to clean up the blood by hand, or lick it up, under threat of the lash. If touching a holy or forbidden animal was degrading to the men of India, touching the blood of humans was the ultimate defilement.

"I wish to show the Natives of India that the punishment inflicted by us for such deeds will be the heaviest, the most revolting to their feelings . . . ," wrote the officer in charge. After each "culprit" had cleaned up his share of the bodies and carnage, he was immediately hanged.

The British could be just as cold-blooded as the mutineers. In September company forces attacked Delhi and massacred every man on sight—not just the combatants fighting British rule. The British slaughtered defenseless citizens in cold blood, striking down some fourteen hundred men in one neighborhood alone.

"It was literally murder," wrote one officer witnessing the killing of Indians in Delhi. "I have seen many bloody and awful sights lately, but such as I witnessed yesterday I pray I never see again. The women were spared, but their screams, on seeing their husbands and sons butchered, were most painful."

The civil war brought to an end the reign of the Mughal emperors who had colluded with the company as proxy rulers. The last emperor, Bahadur Shah, wrote from exile:

My life now gives no ray of light,
I bring no solace to heart or eye;
Out of dust to dust again,
Of no use to anyone am I.
Delhi was once a paradise,
Where Love held sway and reigned;
But its charm lies ravished now
And only ruins remain.

When the uprising was finally put down, Parliament rescinded the privileges of the East India Company in India and revoked its charter. With the stroke of a pen the company ceased to exist. Henceforth, the British Crown would be the government of the subcontinent; Victoria would become empress of India.

In a quarter of a millennium of existence the company had amassed possessions to rival Charlemagne's and created an empire on which the sun never set; it was the first global multinational and the largest corporation history has ever known. Yet it failed spectacularly at one significant task: to govern India in peace. However ingenious the idea and potentially profitable the industry, growing tea in India could not save the Honourable Company from extinction.

18

Tea for the Victorians

B y the time the Chinese realized that Fortune had stolen an inestimable treasure from them, it was many years too late to remediate their loss. His theft helped spread tea to a wider world at lower prices. He democratized a luxury, and the world has been enjoying it ever since.

Like sugar, coffee, tobacco, and opium, tea was among the first mass-produced, mass-marketed global commodities. Though tea would not cause the Industrial Revolution, its popularity in Britain and the increasingly easy access to it brought about by the advent of Indian tea spurred British industrialization along.

These global commodities could almost be considered a form of kinetic economic energy—picked for a penny, sold for a pound, and transported around the world. They rearranged the axes of power all along their supply chain, each step of the way from mountain farms to British homesteads. Through a simple drug, tea, Chinese coolies were connected to American traders and Parsee bankers in Canton, to financiers in London, to mothers and children in Manchester enjoying their breakfast.

Tea likewise revolutionized Britain's capital and banking systems and influenced the rapid growth of trade networks in the

Far East. It was instrumental in extending the reach of British colonialism as the empire expanded to include countries such as Burma, Ceylon, East Africa, and others where tea could be grown, establishing economies in places that were previously considered little more than blighted jungles. Tea also influenced efforts to colonize areas such as the Caribbean and the South Pacific, which could help meet Britain's demand for sugar. The Oriental trade catapulted Britain and the pound sterling to nearly two centuries of global prominence, a feat that no sparsely populated agrarian island nation could otherwise have achieved.

Tea changed the role of China on the world stage. The tea trade gave birth to the colonial territory of Hong Kong, a thriving city—now once again a Chinese city—and the capital of the Orient. Some have argued that Hong Kong demonstrates the role China might have taken on the world stage had a series of tea-trade-inspired revolutions not halted progress on the mainland for much of a century. The foreign presence in China and the havoc wrought by the tea/opium exchange jointly undermined the imperial leadership of the Qing Dynasty. The fall of the Qings, in turn, led to the rise of the Nationalist Kuomintang and eventually of the Chinese Communist Party, a turn of events that would later result in the modern-day division between Taiwan and China. No one can reasonably lay the responsibility for these historical developments on tea alone, but neither can one ignore the role that foreign desire for this quintessentially Chinese commodity played in opening up China to the West and in the country's subsequent fall from imperial self-sufficiency.

Apart from its geopolitical ramifications, the tea trade affected nearly every aspect of the economy.

Transportation

By the 1850s the passage to London from China took one month less than it had even ten years previously, spurred on by the race to bring tea to market. Tea gave rise to the fastest ships under sail; their speeds have never been matched.

For the first two hundred years of the tea trade the only ships making the trip to China out of Britain belonged to the East India Company, the sole business chartered to trade in the Far East. These boats, known as East Indiamen, were slow floating warehouses. The "tea wagons" sailed from the Thames to Canton and back; tea's journey to the Mincing Lane auction in London took as long as nine months after picking and sometimes an entire year. This meant that even the finest grades of tea, the flowery pekoes and souchongs, lost their edge. No "new season" tea was ever available, although travelers' and merchants' reports suggested that the "first cut" off the first blush was the choicest brew. While only a few of London's tea drinkers were ever likely to notice a degradation in quality, those in the tea trade knew there was vast room for improvement, that a premium might well be paid for higher-quality, fresher teas. But without competition, with no challenger attempting to bring tea from hillside to table in the same season, there was little impetus for innovation and no improvement for many years.

The nineteenth century saw tremendous technological advances in shipbuilding. After Britain defeated Napoléon in 1815, there was no longer a pressing need for the old British warships, heavily gunned and self-sustaining enough to remain at sea for long stretches, avoiding land. In the period of peace that followed, ships became longer, sleeker, and faster.

When the East India Company monopoly over China ended in 1834, new trading companies sprang up to claim a portion of the China trade—names still revered in the Orient: Swire, Jardine, and Matheson. These firms fought for a cut of the lucrative tea profits, challenging the East India Company, which like its ships, was bloated and inefficient. As increased competition created new incentives for faster boats, the tall-masted sailing ships became even more refined.

When the 1849 repeal of the British Navigation Laws allowed American-built ships to sail to and from China, the Americans could finally offload Chinese tea right onto British docks—and did so weeks ahead of British-built ships. The American vessels, modeled after the finer hull lines of the swift privateers from the War of 1812, could make the run between New York and Canton in under a hundred days. With speed at a premium, British shipbuilders went back to their drawing boards; they shaved the bows, narrowed the hulls, and raked the masts to compete with the best of Boston's designers.

In a matter of twenty years, these three factors—the end of Napoléon, the end of the East India Company's monopoly in China, and the entrance of the Americans into China shipping—accelerated the delivery of tea and revolutionized navigation under sail. The new ships, called tea clippers, were immediately recognizable by their long, low hulls that had a "fish head" stern hanging sharply over the water. They were square-rigged and triple-masted, "a perfect beauty to every nautical man," as one captain remarked.

With the advent of the clippers, the tea trade also became a hugely popular spectator sport. Once the first raking masts of the Oriental fleet were sighted in the English Channel, the City of London turned out on the banks of the Thames to watch the annual Tea Race, when the China clippers delivered the new season's picking. Tugs were engaged, signals were flashed from every headland all the way up to London, wagering began, and fortunes were made or lost on which ship would be the first to throw a box of tea over its gunwales and onto the wharf. There was as much interest in this race as there was in the results of the derby.

The tea clippers remain the fastest sailing ships in the world, in part because they were marvels of engineering and enterprise, and in part because there was never again a need for a big, fast sailing boat. Trade with the Far East became so valuable that the French undertook the building of the Suez Canal. Although clippers couldn't sail in the waterway—the Red Sea's winds were too challenging—a steamship could reach China in half the time of a clipper. With well-placed fueling stations, the journey to China and India grew ever easier. By 1869, when the Suez Canal was complete, all the improvements in navigation brought on by tea would become a thing of the past. The ambitions of the British merchant fleet could be fueled by reliable and steady coal, not fickle wind.

Manufacturing

Tea's light weight meant that a merchant ship transporting it needed ballast to stay trim, and for most of the early years of the tea trade, that ballast consisted of blue and white Chinese porcelain. Although this merchandise tended to be undervalued by traders, who could make larger profits on more desirable commodities such as silk, it was considered useful as "kentledge," the ballast padding between layers of tea crates, and it had the added benefit of ensuring against leaks when it lined the hull and keel.

Luxury items such as tea were high risk: They were vulnerable to water damage, and a ship was always in jeopardy of being lost at sea, so porcelain helped spread the risk around. Porcelain also protected the more valuable cargo from dirty bilgewater.

Tea's growing consumer base encouraged the development of the porcelain industry in Britain—among the very first industries to take advantage of the nineteenth century's mechanical innovations. Prior to the eighteenth century, no European factories could make a ceramic teacup capable of holding boiling water.

European clay could not meet the service demands of tea the way Chinese clay could, for in Europe clay lacked the essential ingredient of kaolin. Chinese porcelain was fired at a high temperature, so it was cheap and sturdy, with a strong transparent glaze. Because European clays were fired at a lower temperature, they had porous glazes and were more likely to break.

The need for more durable pottery stimulated an industrial race in England. Could British manufacturers make tableware harder and more inexpensively? English factories had been working with stoneware, which was heavy, coarse, and fragile, but by virtue of its lower transportation costs it could compete with Chinese porcelain on price. Ultimately, by about 1750, European factories discovered the secret of porcelain production, and a new industry was born that took advantage of the mechanization that was then transforming British industry. (Ironically, one of the earliest potters to make use of these technological advances was Josiah Wedgwood, who could count among his grandchildren Charles Darwin, the naturalist and contemporary of Robert Fortune.)

The tea trade, through the porcelain trade, also stimulated the notion of China and the East as "exotic" places. The iconic images of chinoiserie—weeping willows and towering pagodas, and

demure women clothed in flowing robes—became familiar as a result of being stamped or painted on the sides of imported teacups. This romanticization of the Orient was highly useful for advancing the imperial project; it made a pretty thing of an unknown place and lent an air of enchantment to what might otherwise have been considered a fearsome, treacherous journey abroad. Great poverty was devastating England, along with disease and dislocation. Cities encroached on the countryside, and its former inhabitants were now working in factories, breathing smog, and living in crowded tenements, but the images visible everywhere on porcelain represented a link to a wider, bigger, and better world. It was a world of trade and possibility, a world that England could conquer.

Life in England

As had been predicted by the Honourable East India Company, Indian-grown tea, aided by the repeal of tea taxes and advances in shipping—and increased competition with the end of the company's own monopoly—led to an overall decrease in tea prices. Cheaper tea also meant that unscrupulous dealers felt less need to pad their product with other plants and dangerous chemicals, so the tea's quality improved. Although England had been a nation of tea drinkers for over a century, cheaper tea became a boon to the rapidly urbanizing country.

Demographers and doctors had long noticed a drop in the mortality rate as the taste for tea became increasingly popular. With the growth of cities in the eighteenth and nineteenth centuries came a rise in levels of pollution and disease. Cholera, which had long plagued the Indian subcontinent, made its first appearance in England in the 1830s when infected sailors, drinking water from ships' barrels filled in India, returned to their home port and spread the deadly bacteria through local sewers. By midcentury, cholera epidemics were repeatedly wiping out Londoners by the tens of thousands; the outbreak of 1848–49 alone claimed fifty thousand lives—all from drinking water.

Countries such as England, where tea was preferred to coffee that was steeped in hot but not boiling water, reaped immediate health benefits from their drinking habits because boiling water killed the microorganisms that spread contagion at close quarters. Even under normal circumstances London's drinking water was far from sanitary, owing to the density of the city's population and lack of proper waste removal. A nation of tea drinkers was more likely than one of coffee drinkers to survive the repeated infestations that were a product of the global economy of the Victorian era.

Tea was a boon to the imperial project as well. It became a standard part of rations for the British army, as well as for native troops in the colonies. As Englishmen were slogging through the jungles of the tropics, tracking the boundaries of empire, they comforted themselves with a cup of tea and simultaneously kept water-borne illnesses at bay.

As previously noted, sugar was another key commodity in the intricate economy of the British Empire, a product of the queen's remaining colonies in the New World: Barbados, Jamaica, and the Virgin Islands. Britain had a glut of sugar, and tea gave Britain somewhere to dump it.

Tea with sugar provided Britons with a convenient source of calories. The urbanization of Britain meant that the poor no longer had easy access to farm products, and while tea was not inherently nutritious, it could be drunk with milk, a protein, and sugar, a cheap and dense source of energy.

Prior to widespread tea drinking, factory workers obtained much of their calorie intake from beer and ale, which made for a less than ideal workforce. Beer drinking could be tolerated in workers doing primarily manual labor, as was the case in preindustrialized economies, but it posed a serious problem in the industrialized sectors of Britain's economy, where fine motor skills were required. A drunk worker was a danger around the fast-moving looms and needles of Manchester's textile industry. But by drinking sugared tea and eating bread, plus meat on Sunday, Britons could get all the calories they needed without the risk of intoxication. Indeed, tea had a stimulant effect; it focused the minds of the workforce, helping them to concentrate better on their demanding jobs.

Fermented drinks such as alcohol also have the benefits of killing parasites and bringing liquid calories to the diet, but by the start of the eighteenth century beer production was eating up nearly half of the wheat harvest in Britain. There was no possible way for Britain's domestic agriculture to feed the rapidly expanding population and keep them in beer, too. There just wasn't enough farmland for every new mouth in the industrial era. Calories had to come from an outside source, one beyond the boundaries of the British Isles, from the wider shores of the empire. The pursuit of food has always shaped the development of society, and in the days of the Victorian empire, the very start of our modern industrialized global food chain, tea with milk and sugar became the answer to Britain's growing need for cheap nutrition.

European countries such as France and Germany that continued to choose alcohol as the staple drink lagged fifty years behind Britain in the process of industrialization.

Other benefits accrued from the choice of tea over beer, particularly for the young. Pregnant women drinking tea rather than beer significantly improved the health of the infant population. Tea also contains antibacterial phenols, plant-based chemicals that act as natural disinfectants. Since it was typical in Britain to breast-feed babies for the first year, mothers opting for tea rather than beer meant British babies were no longer exposed to as much alcohol. Tea reduced infant mortality and gave an immunizing boost to the population when industrialization was demanding an ever larger workforce.

By the mid-nineteenth century, "tea in the drawing room" marked the first time that the ritualized notion of tea drinking took root in British society. Afternoon tea, which had initially been a ceremony in the higher orders of society, became a widespread custom following the gradual drop in prices. Teatime was a period of enjoyment, of visiting, of having a brief chat during the long hours between noon and the evening meal. The industrial revolution, which tea helped spur, created enough excess capital that Britons could finally enjoy the fruits of their affluence.

Today in the West it seems that a new study is released every day examining the health benefits of drinking tea—from its antioxidant and anticarcinogenic properties to its role in stabilizing diabetes, raising metabolic rates, and lowering the risk of obesity or boosting the immune system. Many of these claims might still require further scientific substantiation, but any regular consumer of tea will confirm that tea drinking increases mental alertness and short-term memory and also lowers stress. Experts are examining tea from every direction as a magic elixir for respite, recreation, and a better, longer life.

England's great tea experiment in India was a phenomenal

success. More people now drank more tea for less money. Whithin twenty years of Fortune's theft of Chinese trade secrets, the tea trade shifted away from China to British dominions. When a single species was transported out of its native soil, the world was never again the same.

19

Fortune's Story

To bring Fortune's own story to an end, let us imagine a day at Kew, sometime in the 1870s.

Having sailed down the Thames, Fortune, now in his sixties, arrives at the gardens to pay homage to his botanist friends and colleagues—among them the current director, Joseph Hooker, the man once imprisoned in Darjeeling by the rajah of Sikkim along with Archibald Campbell. Hooker is a great friend of the naturalist who is stirring up the scientific world, Charles Darwin. Fortune has a lovely visit, chatting with other botanists about new discoveries, sharing memories of past scrapes, and counting himself—at last—a man among equals.

Given Fortune's taciturn nature and his ability to keep his own company for years on end, he excuses himself from the distinguished gathering of other botanists after finishing his tea and takes a walk through exquisite gardens that once existed for the pleasure of royalty but are now, rightly, a hothouse of scientific development.

Fortune steps up to a great greenhouse, the Palm House, gloriously situated on a hill. Kew's first director and Fortune's former superior, the father of the current director, William Jackson

Hooker, had the Palm House constructed with a roof 66 feet high to accommodate the great height of its eponymous tree, making the structure the focus of Kew. The tropical giants grow until they reach the ceiling. The all-glass Palm House utilizes the fundamental principles of Ward's discovery, and its engineering is based on developments in shipbuilding: It is essentially an upside-down glass ship's hull. Nestled amid the palm trunks are further examples of tropical imperial exploration—spices, fruits, timber, fibers, perfumes, and materia medica, the prizes of previous plant hunters.

Fortune wanders on, to a part of the garden that might have been his very favorite: the Chinese pagoda. The pagoda was built in 1761 for a dowager princess in the midst of a fad in Britain for anything exotic and Chinese. A Chinese influence has become popular in decorative arts throughout the country: in the furniture designs of Thomas Chippendale, in textiles, in fine porcelain, and in the Kew pagoda built of English redbrick. Fortune is reminded of China in everything he looks upon. And the pagoda, more than anything else, recalls his long years of travel and the images of a faraway country that still haunt him decades later.

In truth, we have no record of what Fortune experienced on this visit or precisely when it occurred. What we do know of Fortune's adventures comes from his published works on China, India, and Japan, and from the copious records kept by the faceless, dutiful, admirable East India Company clerks. Although a great deal of the company's archives were destroyed when the charter was revoked after the mutiny, there is, thankfully, an ample supply of documents in the British Library. However, none of Fortune's private papers survive except his letters to others, such as those to Joseph Hooker, director of Kew.

Following his first tea trip, Fortune returned home to his wife and children—but only very briefly. He scarcely had time to write his memoirs and reacquaint himself with his loved ones, see to the younger Fortunes' well-being, and impregnate his wife before the company called again.

Fortune's discovery of the coloring of green tea prompted a change in the tastes of the British public. The unveiling of the chemical greening of green tea at the Great Exhibition of 1851 marked a turning point: Britons now wanted their tea black and only black. The tea makers Fortune sent to the Himalayas were versed in the art of green tea, but the results of his findings about chemical dyes so altered British tastes, he was sent out to China for a third time with explicit instructions to hire black tea experts.

Fortune's next trip to China also included another form of espionage when he became a drug smuggler. While Britain engaged in thievery to build its own tea trade, China, too, was plotting its botanical thievery: It planned to raise a domestic opium crop to compete with Indian Patna-raised opium. Just as tea could find a second home on the other side of the Himalayas, *Papaver somniferum*, the opium poppy, could be transplanted to the lush rolling hills of China.

In the years after the First Opium War, although Britain had won the right to deal opium in China, China's native opium was busily supplanting this trade. China had exercised all its ingenuity on discovering less expensive ways to get its fix. The company, a dying multinational in its last throes, feared that China's own domestic crop would one day squeeze it out of this lucrative market entirely. Since it had a prominent horticulturist already working undercover for it there, what could be easier than expanding his remit and asking him to obtain for the company as much information as possible about China's nascent opium crop?

Fortune returned to the company the following high-risk and ultimately rewarding items:

> 2 dried specimens of the poppy from which Chinese
>   opium is made
> 1 opium knife used by the natives in collecting the
>   poppy seed
> 1 paper of seeds of some poppy

"The seeds and specimens I send now will give the means of ascertaining . . . what those differences [between the Indian and Chinese varieties] are and whether they account for the difference in quality between Indian and Chinese Opium," Fortune wrote to the company. Botanists in India soon studied what exactly those differences were. In his popular and scholarly published works, Fortune made no mention whatsoever of his opium investigations.

When the Indian Mutiny ruined the East India Company, Fortune found himself a new employer for a brief time: the United States government. By now his fame was secure, and his accomplishments were widely celebrated. He was the father of a whole new industry in India, and the United States wanted its own share of the potential wealth. It had been taken for granted that America was a coffee-drinking culture ever since the Boston Tea Party, when a gang of patriots dressed as Mohawk Indians threw an entire shipment of East India Company tea into Boston Harbor to protest the high taxes imposed by the British Crown. But nearly a century later the American government had, in fact, ambitious plans for tea. It was believed that the agricultural regions of the South would make the United States a viable competitor in the lucrative world tea economy. Fortune was asked to assess whether or not tea could flourish in the hilly and humid southern Appalachian states, the Carolinas or Virginia. Labor was still cheap in the South.

In 1857, the U.S. Patent Office hired Fortune to bring tea seeds to America, paying his standing rate: £500 a year and all expenses. Caught up in the frenzy of industrialization, the Patent Office believed American engineers could harness steam power to automate the processing of tea.

Fortune sailed for China again in March 1858, arriving in the Chinese interior in August. By December he had ready two Wardian cases of tea seeds and plants for his American employers. After many years of trial and error he had become an expert tea hunter.

As soon as the cases arrived in Washington, the commissioner of U.S. patents unceremoniously fired Fortune, presumably in the belief that American gardeners could take care of the by now flourishing seedlings. Indeed, the tea thrived in the Patent Office's gardens. By 1859 some thirty thousand well-rooted plants from Fortune's shipment were available for distribution to plantations in the South. By 1860 the dispersal of tea seeds and plants had become a significant portion of the work of the Patent Office's agricultural division.

When the Civil War broke out in 1861, the newly formed U.S. Department of Agriculture lost all communication with the areas involved in the tea trials. By the war's end, without slave labor American tea could not compete with the lower overheads in Asia: The price of picking a pound of tea in the United States was six times that of tea picked in China. Although there were a few more attempts at establishing an American tea industry, it died stillborn. Fortune spent much of the Civil War trying to recoup his fee from the Patent Office—which, it seems, he never accomplished.

His final trip to the Far East, in 1862, took him to both China and Japan and was the only one he made as a private citizen, working for nursery firms and paying his own way. Fortune could finally keep for himself a share of each and every discovery and all ensuing profits. Japan, a previously closed country with a latitude similar to Britain's, was thought to possess botanical prizes that would adapt nicely to the climate of England. Fortune benefited handsomely from his finds there; his botanical discoveries were embraced by enthusiastic plant collectors throughout Britain, and so in his final years he became very wealthy. Moreover, after having spent so much time in Asia, meeting mandarins and farmers, peasants and poets, he had developed a good eye for the arts and decorative objects of the East, which remained highly prized by the Western aristocratic and merchant classes. When he died in 1880, his estate was valued at over £40,000 (at least $5 million today).

Fortune popularized a remarkable variety of flora in the wake of his Chinese travels. The plants he "discovered" in the Orient number in the hundreds and include the bleeding heart, the winter jasmine, the white wisteria, twelve species of rhododendron, and the chrysanthemum. He corrected Linnaeus's definition of tea's taxa by proving that green and black teas were one and the same, and he improved the health of Britain by revealing that the Chinese were coloring green tea with poison. Fortune's experiments with transporting tea seeds in Wardian cases also made it possible for many plants, inluding the great trees of England—the towering oak and chestnut—to migrate. Prior to his improvements on the Wardian box, many species could not be replanted in the colonies because acorns and chestnuts, much like tea seeds, simply did not travel well. His work made it possible for entire agricultural economies to find new markets in new homes.

But Fortune's world of plant hunting was already receding quickly into the past. Once the Suez Canal opened, ships were able to travel between China and England in little over a month, avoiding the treacherous temperature changes of traveling around the Horn and minimizing the threat to plant cargo. Telegraph cables were wired from one part of the globe to another so that the improvisational bravado that marked his time in China faded away as information became more easily and widely disseminated.

It might have seemed to Fortune that all the grandeur of the earth had been completely cataloged and numbered. There was now new work to be done utilizing Charles Darwin's theories of evolution. Natural history was no longer the mere act of classification; it became a hunt for the narrative of how species evolve adaptive characteristics through competition and natural selection.

There would be no more Robert Fortunes: The East India Company was gone and with it the institutional need for plant-hunting botanists. There was no longer any giant corporate monopoly willing to pay for research and development on the scale of the Honourable Company. Fortune had played his part on a grand stage as a younger man. In his time he had done his part to enrich Britain; he made it more green and pleasant still. So many of Fortune's precious seedlings, so delicately packed and cared for in faraway Cathay, had found a home at Kew, under the shadow of the great pagoda.

By the time Fortune was an old man, the tea plantations of India had outstripped those of China, whose tea would never again be as competitive in the Western market. Fortune's tea theft would continue to reap benefits: Tea would be brought to Ceylon, to Kenya, to Turkey, and it would be judged to be good and even excellent quality. When Robert Fortune stole tea from China, it was the greatest theft of protected trade secrets that the world has ever known. His actions would today be described as industrial espionage, viewed in the same light as if he had stolen the formula for Coca-Cola. Any number of international treaties now police foreign exploitation and protect national commercial treasures.

Today there is only guarded enthusiasm for the mass globalization of indigenous plant life. We know now that when species are brought to new habitats, where they may have no natural predators or competitors, they can overpopulate and decimate local ecosystems. Entire islands have been overrun as the result of the kind of botanical frontiersmanship that Fortune and his contemporaries routinely practiced.

Robert Fortune died in 1880. Little is known about how he spent the very last years of his life. For reasons of her own, his wife, Jane, burned all his papers and personal effects upon his death.

Acknowledgments

First thanks belong to my agent, Joy Tutela, who with grace and guidance helped me broaden a story about a gardener into a book about his world. Joy has fought for me, nurtured me, and put me in my place ever so gently. She and her colleagues, David Black, Susan Raihofer, Leigh Ann Eliseo, Gary Morris, Johnathan Wilber, Caspian Dennis, and Abner Stein, are the noblest, kindest, and, dare I say, the best agents in the entire world.

Warmest thanks to my editor at Hutchison, Paul Sidey, whose enthusiasm for Robert Fortune frequently outpaced my own, and to James Nightingale, Laura Mell, and Emma Mitchell. Thanks to the entire Viking team: Rick Kot, Holly Watson, Laura Tisdel, Meghan Fallon, Kate Griggs, Nancy Resnick, Jerry Buckley, Margaret Payette, and Paul Buckley.

Lord, make me worthy of my friends:

Weidong Fu is my honored teacher, translator, and companion whom I miss terribly. I can only hope this book is a fitting souvenir of our travels together.

Scott Anderson brought me Robert Fortune one very dark winter and, with bottomless patience and generosity, weathered my tantrums. He read every word of the text many times over and then told me, quietly, to make it better. I would not be the writer I am without him. Moe, Donna, and Kelly Anderson taught me to love gardens. Donna died before she could read this book, which owes a tremendous debt to her delight and skills in the garden. Every tomato I eat reminds me of her.

Kim Binsted is my fellow traveler. It is entirely her fault that I ever lived in China. She welcomes me to Hawaii each winter, no matter how tragic I become. I will follow her anywhere. Even to Mars.

Joel Derfner is the most talented man alive. A girl could not find a sexier or funnier companion on the terrible road to publication.

My first readers, Rachel Elkin Lebwohl, Victor Wishna, Saul Austerlitz, and Joel Derfner, provided my first moments of reward in three long years—in addition to sterling line edits.

Megan Von Behren was my genealogist, and Daniel Von Behren was my cheerleader—and I am thrilled to be their wife.

Much of this book was researched and written far from home, and I am grateful to those who gave me shelter and friendship in beautiful places. In London: Julian Land, Miriam Nabarro, and Coco Campbell; in Tuscany: Henry, Tory, Elizabeth, and Joe Asch; in Maine: Marc, Lauren, David, and Delia Laitin; in Delhi: Janaki Bahadur and Christopher Kremmer.

Other friends of Sarahworld who deserve a kiss and baked goods: The Magnificent Seven who kept the wine flowing, Karen Bekker, Catharine Clark, Tammy Hepps, Charlie Paradise, Adina Rosenthal, and Shana Sisk. I have been blessed with wonderful teachers: Darlene McCampbell, Earl Bell, Richard Ford, and Alan Richman. Chicago would never be warm without Maria, Sergio and Pierralberto Deganello, Tino Palacio, Mindy Graham, Rosie Humphrey, and Burt Friedman. Hawaii would be warm but no fun without Jennifer Baker, Jason "Big Red" Jestice, and Walter Eccles. Charles Coxe gave me assignments when I needed them most. Laila, Talia, and Mia Veissid and their parents, Marco and Phyllis, cheered me through dreary revisions and continue to do so.

Special thanks to Barney Rose, Clare Hollingworth, Betsy and Jeff Garfield, Evan Cornog and Lauren McCollester, Stephanie Jordan and Adam Brown, Shai Ingber, Elizabeth Hamilton, Jason, Beth, Barbara, and Bill Myers, Richard Bradley, Elaine Land, Anita Fore and the Author's Guild, and the late Gladys Kenner, who always said China was no place for a young lady.

Last, tea and I share a birthday: The first flush picking is celebrated on the third day of the third lunar month, which corresponds to the third of April (or thereabouts). If I have expressed that happy birthright, it is in no small part thanks to my parents, Helen Cohen Rose and Gerald Rose, without whom this book (and I) might never have flowered.

Notes

As this is a work of popular history, not a scholarly undertaking, I have avoided the use of footnotes and tried to steer clear of mentioning sources in the body of the text. Nevertheless, this is a work of nonfiction, and anything in quotes comes from a letter, memoir, newspaper, or other contemporaneous source.

I have relied heavily on Robert Fortune's four memoirs, his letters to the East India Company, and other company documents housed in the British Library. Over five hundred books and documents were consulted in putting this project together. Each of the broader themes—Fortune, the East India Company, China, and professional botany—have been the subject of inquiry by minds greater and more learned than my own.

The exhaustive work on tea is William H. Ukers's aptly named *All About Tea* (The Tea and Coffee Trade Journal Company, 1935). For a more narrative tale there are several excellent popular histories, but I most enjoyed the spirited *The Great Tea Venture* by James Maurice Scott (Dutton, 1965).

It surprised me how much fun it was to explore the world of the naturalists and botanists in Victorian England and the empire, and I can't find enough praise for the scholars who got there first: David E. Allen's *The Naturalist in Britain: A Social History* (Princeton University Press, 1994), David Arnold's *The Tropics and the Traveling Gaze: India, Landscape, and Science, 1800–1856* (University of Washington Press, 2006), Fa-ti Fan's *British Naturalists in Qing China: Science, Empire, and Cultural Encounter* (Harvard University Press, 2004), Sidney W. Mintz's spectacular *Sweetness and Power: The Place of Sugar in Modern History* (Penguin Books, 1986), Donal McCracken's *Gardens of Empire* (Leicester University Press, 1997), Henry Hobhouse's *Seeds of Change* (Harper and Row, 1986), and Sandra Knapp's *Plant Discoveries: A Botanist's Voyage Through Plant Exploration* (Firefly Books, 2003) were all tremendous resources. I am indebted to Sue Minter's *The Apothecaries' Garden* (Sutton, 2000) as well as the pamphlets published by the Chelsea Physic Garden and to the garden itself— among London's most splendid treasures.

As any student of China might, I depended on masters to instruct me in Qing history: Jonathan Spence, John King Fairbank, Immanuel C. Y. Hsu, Philip Kuhn, and Fredric Wakeman all wrote brilliant and breathtaking works of expansive history that I eagerly devoured. For further reading on the Taiping Heavenly Kingdom, I recommend Spence's *God's Chinese Son* (W. W. Norton, 1996). On opium, Zheng Yangwen's *The Social Life of Opium in China: A History of Consumption from the Fifteenth to the Twentieth Century* (Cambridge University Press, 2005) is excellent. For life in rural China, Nancy Berliner and the Peabody Essex Museum's *Yin Yu Tang: The Architecture and Daily Life of a Chinese House* (Tuttle Publishing, 2003) is a generous resource.

I fear there is less compelling historical work on the East India Company than there should be. No text captured the grandeur or complexity of the company in a way that satisfied me as a researcher. Nevertheless, I leaned on several solid works, among them Peter Ward Fay's *The Opium War* (University of North Carolina Press, 1975), Nick Robbins's *The Corporation That Changed the World* (Pluto Press, 2006), John Keay's *The Honourable Company: A History of the English East India Company* (HarperCollins, 1993), Patrick Tuck's *The East India Company* (reprinted by Routledge, 1998), H. A. Antrobus's *A History of the Assam Company, 1839–1953* (privately printed by T. and A. Constable, 1957), as well as David Macgregor's *The Tea Clippers: An Account of the China Tea Trade and of Some of the British Sailing Ships Engaged in It from 1849 to 1869* (P. Marshall, 1952).

In learning about Fortune, China, the East India Company, and tea, I benefited from the able advice of tea experts Mr. Lu Shun Yong, his partners, Mr. Wang and Mr. Shi, as well as Alan Stokes and Michael Harney; China scholars Carsey Yee, Jan Wong, Hayes Moore, and Susan Thurin; the peerless Richard Morel and the staff at the British Library Asia collection, and the entire University of Chicago library system—my favorite library on earth.

Beijing was previously known as Peking to English speakers. In 1949 the Communist government replaced the anglicized Peking with the official pinyin, Beijing, but it did not come into common usage until the 1980s when the government began using Beijing on official documents. Many of the older China hands whom I first met during the handover of Hong Kong still use Peking to refer to the capital. It retains a kind of colonialist hue that I decided to keep throughout the text, as anachronistic as it is.

Fortune was often the first or among the first Westerners to go into a rural area, and it is not always clear what location or object he refers to in Chinese because he traveled at a time when there was no systematized Chinese to English transliteration. Where possible I have strived to include the pinyin. But I am not, alas, a Chinese speaker, and I was dependent on translators throughout. All mistakes, of course, are mine.

Index